LIFE IN AND AROUND
FRESHWATER WETLANDS

By the same author:

Life In and Around the Salt Marshes

LIFE IN AND AROUND FRESHWATER WETLANDS

A Handbook of Plant and Animal Life In and Around Marshes, Bogs, and Swamps of Temperate North America East of the Mississippi

by MICHAEL J. URSIN
With illustrations by the author

Thomas Y. Crowell Company   New York   Established 1834

The Introduction to this book, by John H. Mitchell, originally appeared in *Massachusetts Audubon*. It is used here by permission of the author.

Library of Congress Cataloging in Publication Data

Ursin, Michael J
    Life in and around freshwater wetlands.

    Bibliography: p.
    1. Wetland ecology—North America. I. Title.
QH102.U77   574.5'2632   74-13632

ISBN 0-690-00673-X
    0-8152-0378-0 (Apollo pbk.)

1 2 3 4 5 6 7 8 9 10

To those organizations and individuals
who have worked so hard to protect
and preserve our valuable wetlands

ACKNOWLEDGMENTS

This book has been made possible through the generous assistance of many. I wish to thank in particular Virginia A. Barber, for her typing and editing; Virginia M. Ursin, for her contributions in research; John E. Swedberg, for outstanding color photography; and the Thomas Y. Crowell Company, for their help and patience.

PREFACE

This book is a brief guide to plant and animal life commonly
associated with marshlike freshwater environments. It does not
include every species; rather it describes many common varieties and
representative forms of plant and animal groups.

The foremost purpose of this work is to stimulate interest in our
valuable wetlands.

Taxonomy, or classification, of the plants and animals is utilized at a
minimum, and then only to assist those who wish to investigate
further. Although it is possible to find different classifications for some
species, the system used here is commonly accepted and easily
followed.

CONTENTS

LIFE IN AND AROUND FRESHWATER WETLANDS

INTRODUCTION

by John H. Mitchell
Massachusetts Audubon Society

SOMEWHERE BETWEEN eleven and twelve thousand years ago, the great ice sheet that covered the northern half of the North American continent began a gradual retreat. As it slowly melted back from its southernmost point, it left behind a legacy of lakes and ponds on the channeled and sculptured landscape.

Some of these bodies of water formed in basins that the glacier had gouged out of the bedrock during its advance. Others developed in dips and depressions in the mixture of impermeable mineral matter known as till, and still others collected within the vast clay-lined beds created in the sand and gravel that washed out from the glacier as it melted, or were formed from huge chunks of ice that were left isolated from the body of the ice sheet.

Even before the last remnants of the glacier had melted from the valleys, the myriad lakes and ponds that were left strewn across the landscape began to evolve into freshwater swamps, marshes, and bogs. Most of these wetlands today are found along the floodplains of rivers and streams or in other spots of low elevation. Some, however, because of the sealing nature of bedrock and till, may be found perched high on the sides of hills, or even on mountaintops.

No matter what their origin or location, bodies of open water eventually become wetlands because all lakes and ponds are transitory, remaining open no longer than it takes geological and biological forces to transform them. Part of this evolution is caused by the ability of running water to carry a load of sediment. Over the years, streams feeding into ponds and lakes deposit their baggage of debris on the bottom so that the water becomes shallower. At the same time, in the early stages of succession any outlet is constantly deepening itself through erosion, causing the water to drain off more quickly.

Once the water is shallow enough to admit light to the bottom, plant life begins to speed the process. Submerged vegetation begins to grow along the bottom, and emergent water-tolerant plants start to creep toward the center from the shallow edges. As plants grow and die back, over the years they add layer after layer of decaying organic material until eventually the open water has evolved into what is loosely termed a wetland.

In general, wetlands can be separated into three commonly recognized types, identifiable by their vegetation, each of which represents a former pond at a particular stage of death. If the predominant vegetation consists of grasses, sedges, or rushes and there are no trees, the wetland is a marsh. If there are trees and shrubs, it is a swamp; and if the area is covered with a thick layer of spongy sphagnum moss and is filled with slow-flowing or stagnating water, the wetland is a bog.

Finer classification is more complicated, because as they evolve wetlands can exhibit the characteristics of all these types at the same time. However, generally the progress from open water to dry land follows an orderly succession in which the sun-loving water-tolerant plants are followed by shrubs that can tolerate the shallower conditions, which in turn are followed by trees. In other words, ponds become marshes, marshes become swamps, and swamps become forests. In between lie the varying stages—fresh meadows, with their shallow standing water, deeper fresh marshes, and deep-water shrub swamps. And apart from them all, following laws of its own, is the unique and botanically rich sphagnum bog.

Aside from common species, some unusually interesting forms of plant life are found in wetlands because of the peculiar conditions of moisture and soil. This is especially true of bogs, where, because of the acidic environment and the lack of certain nutrients, plants have

2

evolved specialized mechanisms for survival. Most famous of these are the insectivorous sundews (*Droseraceae*) and the pitcher plants (*Sarraceniaceae*), which gather their necessary nitrogen by trapping and digesting insects. Equally famous for their beauty, if not for unique botanical characteristics, are the varied species of orchids found only in bogs.

It is not only the plants which make wetlands a treasury of life. One of the reasons that conservationists became interested in wetlands is because these areas are excellent feeding and nesting places for many species of waterfowl and other water birds. These include not only the commonly seen black ducks, wood ducks, mallards, and Canada geese, but also the less obvious species such as bitterns, rails, and woodcock. Smaller species of land birds as well nest and feed there. Wetlands are also famous for mammals, especially the commercially exploited ones such as muskrat, mink, and beaver, and they are valued by mammalogists for unique species such as the bog lemming and the water shrew.

However, while each type of wetland provides the requirements for the particular species of plants and animals that live there, perhaps their greatest biological value lies in the fact that as a group, wetlands help to create communities in which a great variety of living things can survive. Biologists have agreed that more diverse communities have a greater ability to remain stable under adverse conditions. This would hold true for those areas which because of their geological make-up have a rich legacy of swamps, marshes, or bogs and thus a great diversity of habitat.

Man has always had a rather ambivalent view of wetlands. On the one hand, wherever possible he has exploited them for their resources. For example, peat was dug from bogs for fuel and medicinal uses, and is still used for horticultural purposes; again, in the past the open freshwater meadows were valued for their forage grasses—although marsh haying was much more common in coastal wetlands. More importantly from a historical point of view, the beavers that were trapped in the wetlands of North America played a key role in the exploration and foundation of the United States and Canada.

But there is also an irrational aspect of man's regard for wetlands. Bogs and swamps were traditionally the dwelling places of spirits and demons. Starting with the epic *Beowulf*, in which the monster Grendel

3

emerges from the fens to ravage the surrounding countryside, English literature and folktales have made constant allusion to the ominous nature of wetlands. Even the language reflects this attitude; the word "heathen," for example, was used to describe brutish, non-Christian people who dwelt on the heaths and moors.

Considering this background, the primal fear that must have struck those first English colonists meeting that dark wall of the wooded swamps of North America can only be imagined. It is not surprising that they took it upon themselves to drain the wetlands and reshape the face of the East Coast. As a result of this cultural heritage, it has been estimated that fifty percent of the original wetlands of New England have been destroyed.

In Massachusetts, those remaining wetlands make up approximately six percent of the total land area. While this may not seem a significant amount, it should be borne in mind that in many communities these wetlands are the last vestiges of open space. In spite of this, and maybe even because of it, a large number are now coming under pressure for development. And this is happening at a time when we are just beginning to understand the importance of their role in many of the natural systems that sustain man.

One of the clearest examples of this role is the relationship of wetlands to flood control. Because of the nature of their soils and vegetation, swamps, marshes, and bogs have a tremendous ability to absorb and retain the excess water that accumulates during rainy periods. As the weather dries, this stored water is slowly released to the surrounding environment, so that streams and rivers continue to run during dry seasons and a relatively stable water flow is maintained. When these wetlands are drained and filled, and the soil and vegetation replaced with less absorbent material, this storage no longer takes place. The surface water runs off quickly and flooding is likely to occur.

This same storage ability also has an effect on the groundwater supply, since surface water tends to collect in wetlands and slowly seep down to the underground reservoirs. This is especially true—and especially beneficial—in those wetlands underlain by highly permeable sand and gravel deposits, because the groundwater in such beds is readily available for human use. When these areas are built upon, the amount of surface water fed into the underground reservoirs is decreased, and the groundwater supply of the community suffers.

Interestingly enough, a recent study of freshwater wetlands in Massachusetts has found that half of those surveyed are located upon these deposits.

There is a potentially even more serious problem when wetlands which are in contact with groundwater are developed. Any wastes discharged into the water-saturated underground deposits are unable to purify themselves through the normal biological processes that take place in septic systems. As a result, the wastes can pollute the groundwater, and the water supply, of any local communities that depend on it. Unlike rivers and streams, the ability of groundwater to cleanse itself is very poor or nonexistent.

Ironically, when they remain intact, wetlands themselves play an important role in the cleansing of both surface and underground water. Pollutants such as sewage, chemical fertilizers, and even pesticides and heavy metals, are trapped and bound into the biological systems of swamps, marshes, and bogs. In fact, it has been suggested that one of the greatest values of wetlands is this ability to act as natural, cost-free sewage treatment plants.

This is especially true in the case of pollution from excess nitrates. When too much nitrate is taken into the human body, it can cause brain damage in infants and serious illness in adults. In recent years, there has been an unnatural input of these chemicals into the environment from such sources as automobile exhaust and fertilizers, and in some areas nitrates have found their way into the water supply of local communities. It is believed, however, that the bacteria living in the oxygen-poor soils of wetlands can transform the nitrate compounds into harmless nitrogen gas. In other words, the wetlands perform a free service which in the human community can only be accomplished by sophisticated and expensive tertiary sewage treatment.

This brings up an interesting question. It is possible that the most important aspects of the role of wetlands have yet to be discovered. As with so many of our environmental problems, we are just now beginning to understand the depths of our mistakes in dealing with wetlands, and we are discovering them backwards. That is, we are finding out the importance of wetlands because in those areas where they have been disturbed, the services that they once provided are no longer operating.

Nowhere is the value of these services more important and yet more poorly understood than in the relationship of man to the landscape. Our environment, as well as our culture, is in the process of monotonizing itself, so that all places seem the same. Few would argue that this has any direct effect on the physical health of the population. However, nothing is known about the effect on the general welfare—the mental state of the culture as a whole. There is something about the deep-shadowed summer swamps and the open expanses of shrub-dotted marshes that goes beyond the practicality of wetlands. There is some undefined and often unrecognized element that provides a break in the monotony of the suburban scene and offers a wealth which is difficult, if not impossible, to measure. Without our wetlands, there is no landscape—only living space.

EXPLANATORY NOTE

This book by no means includes *all* the plants and animals of the wetlands; rather it describes the commonest and most interesting varieties. The individual species are presented as in the example here.

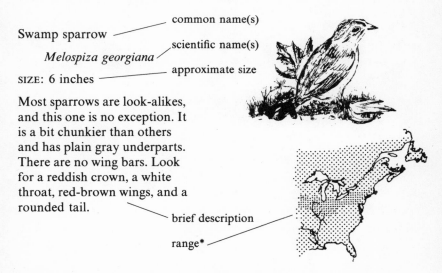

Swamp sparrow — common name(s)

Melospiza georgiana — scientific name(s)

SIZE: 6 inches — approximate size

Most sparrows are look-alikes, and this one is no exception. It is a bit chunkier than others and has plain gray underparts. There are no wing bars. Look for a reddish crown, a white throat, red-brown wings, and a rounded tail.

brief description

range*

---

* When two ranges are indicated on a map, as for certain birds, the upper range represents the summer habitat, the lower one the winter habitat. For species whose seasonal ranges overlap, the area of overlap is shown by heavy stippling.

THE DISTRIBUTION OF FRESHWATER WETLANDS

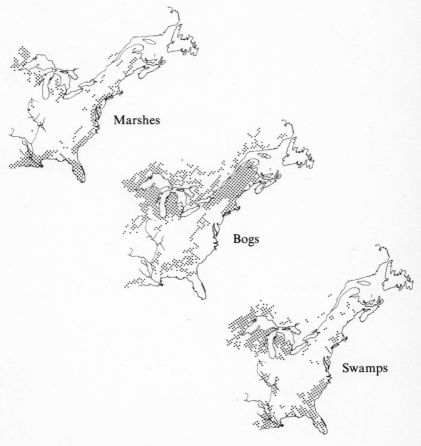

Marshes

Bogs

Swamps

PRIMITIVE FRESHWATER LIFE

Kingdom MONERA—MONERANS
Phylum CYANOPHYTA—Blue-green algae

Kingdom PROTISTA—PROTISTS
Phylum CHRYSOPHYTA—Diatoms (Class *Bacillariophyceae*)
Phylum SARCODINA—Amoebae
Phylum MASTIGOPHORA—Flagellates
Phylum EUGLENOPHYTA—Euglenoids
Phylum CILIOPHORA—Paramecia
Phylum ROTATORIA—Rotifers

MONERANS

The monerans are among the simplest life. They include unicellular forms (some aggregate into filaments), which swim by simple flagellae or glide, or are nonmotile. Bacteria, rickettsiae, and the blue-green algae **(Phylum CYANOPHYTA)** are representative of this kingdom.

The individual organisms are too small to be seen without magnification. Heavy growths of them, however, may discolor or cloud water and may create distinct odors as well.

11

The monerans are the foundation of the food chain. They serve as a direct source of food for higher animal life as well as stabilize the environment through chemical activity.

BLUE-GREEN ALGAE—Phylum CYANOPHYTA

Oscillatoria sp. (×400)

Colonies of *Oscillatoria sp.* form purple masses, sometimes attached to stones.

Rivularia sp. (×350)

This alga forms brown clumps on other plants.

Nostoc sp. (×350)

Nostoc sp. grows in colonies enclosed in a gelatinous mass. The organisms float freely or are attached to objects in the water or to moist shores.

PROTISTS

The protists include single-celled organisms with a distinct nucleus. They are either photosynthetic (that is, containing chlorophyll) simple plants or nonphotosynthetic one-celled animals (protozoans); some forms exhibit both plant and animal characteristics. Protists play a key role in the ecology of freshwater wetlands, furnishing a major food source for small aquatic animals and providing chemical balance.

Diatoms, microscopic algae **(Phylum CHRYSOPHYTA)**, are common in fresh water. They are staple food for many small invertebrates and upon death serve as an important soil builder, forming diatomaceous earth.

A drop of water from a marsh, swamp, or bog when placed under a microscope will reveal a multitude of creatures scurrying, spinning, or flowing in and out of the field of view. These are the miniature inhabitants of the aquatic world called protozoans.

Until recently, any organism that could move from one place to another was classified in the animal kingdom. However, motility is no longer considered in itself a sufficient criterion for placement in a major taxonomic group. The amoebae **(Phylum SARCODINA)** and the paramecia **(Phylum CILIOPHORA)** are the most familiar species of protozoans.

DIATOMS (Class *Bacillariophyceae*)—**Phylum CHRYSOPHYTA**

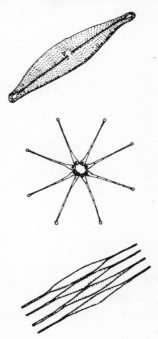

Navicula sp. ($\times$450)

Elongated (pennate) in shape and free-floating (planktonic), this diatom is an important link in the food chain.

Asterionella sp. ($\times$140)

This attractive planktont flourishes in ponds, lakes, and slow streams.

Fragillaria sp. ($\times$200)

Fragillaria sp. is common in still water. It sometimes grows to such abundance that it forms a thick brown scum on the surface.

AMOEBAE—Phylum SARCODINA

Amoeba proteus

This is a simple one-celled animal. It moves by means of temporary extensions (pseudopodia) of the protoplasm. Along with the simple plants, amoebae form the plankton that is a basic link in the food chain.

PARAMECIA—Phylum CILIOPHORA

Paramecium sp.

This is the well-known microscopic "slipper animalcule," common in fresh water.

FLAGELLATES—Phylum MASTIGOPHORA

Ceratium sp. ($\times 275$)

Ceratium sp. is a flagellate—it is equipped with flagellae, whiplike structures used for locomotion.

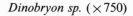

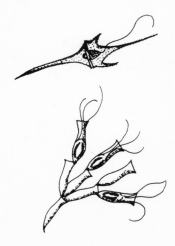

Dinobryon sp. ($\times 750$)

Dinobryon sp. is a microscopic flagellate protozoan with hornlike structures that grow in chains. It is found in pond water.

14

EUGLENOIDS—Phylum EUGLENOPHYTA

Euglena sp. (× 500)

Euglena sp. contains
chlorophyll and is bright
green. It usually is found
among colonies of algae.

ROTIFERS—Phylum ROTATORIA

Polyarthra platyptera (× 200)

This tiny organism, which
spins furiously like a top, is
known as a "wheel
animalcule." It feeds on a
variety of plant and animal life
and is in turn an important
food for other animals.
Rotifers secrete a gelatinous
envelope to survive dry
periods.

PLANT LIFE IN THE FRESHWATER WETLANDS

Kingdom PLANTAE—PLANTS

Phylum CHLOROPHYTA—Green algae
Phylum BRYOPHYTA—Mosses and liverworts
Phylum TRACHEOPHYTA—Vascular plants
  Subphylum SPHENOPHYTINA—Horsetails
  Subphylum PTEROPHYTINA—Ferns, gymnosperms, and
                  flowering plants
    Class *Filicineae*—Ferns
    Class *Angiospermae*—Flowering plants
    Class *Coniferinae*—Conifers

Plants are the most obvious life form in the wetlands. They range in size from microscopic (the green algae) to gargantuan. Dying vegetation may indicate a dying wetland, healthy plants a healthy environment.

Plant growth provides cover and protection for animals and smaller plants, as well as food for animals. Decaying vegetation is a source of enriching nitrogen compounds.

GREEN ALGAE—Phylum CHLOROPHYTA

The simplest representatives of the plant kingdom are the green algae. Unlike seawater, where algae are a predominant form of plant life (seaweeds), fresh water supports only a modest amount of these larger algae. They are usually grass-green or yellowish green; a few are brown or bluish green. Green algae serve as food for many aquatic animals.

Cladophora sp. ($\times 100$)

Pediastrum sp. ($\times 450$)

Scenedesmus sp. ($\times 700$)

MOSSES, LIVERWORTS—Phylum BRYOPHYTA

Mosses and liverworts are more complex plants than algae, having several layers of cells and a protective "skin," or epidermis. They must be associated with water at some time during their life cycle to ensure survival and reproduction.

Water moss

Dichelyma capillaceum

SIZE: Up to 3 inches

D. capillaceum is found in tangled masses on submerged objects. It provides cover for a variety of small animals. Water moss prefers a cooler climate.

18

Boatleaved sphagnum

   Sphagnum sp.

SIZE: Forms clusters up to 1 inch

Sphagnum moss grows in
dense green mats over the
surfaces of bogs, creating the
"quaking bog." These mosses
store water in their cells which
is released during dry periods,
thereby preserving the
valuable wetlands moisture.

Common moss

   Philonotis sp.

SIZE: 2 to 3 inches

Philonotis sp. forms a mat on
rocks and shores of pools and
bogs. When dead the plants
turn white.

Liverwort

   Riccia sp.

SIZE: 1 inch

   Ricciocarpus sp.

SIZE: ½ inch

Riccia sp. floats beneath the
surface of the water. It may
grow in dense masses and wash
up on the shore in green clumps.

Ricciocarpus sp. is distinguished
by its purple-green color. The
underside of the plant has
scalelike red-purple roots.

VASCULAR PLANTS

The vascular plants **(Phylum TRACHEOPHYTA)** make up the bulk of familiar wetlands vegetation. The phylum is comprised of a vast number of plant varieties having complex structures such as leaves, roots, stems, flowers, and seeds—there are in excess of 260,000 species. They provide food, cover, and soil and water stabilization.

HORSETAILS—Subphylum SPHENOPHYTINA

Water horsetail, water pipes

Equisetum fluviatile

SIZE: 2 to 5 feet

This is a primitive plant that grows in shallow water and mud banks. The leaves develop in whorls at the stem joints.

Ferns, Gymnosperms, Flowering Plants
(Subphylum PTEROPHYTINA)

FERNS—Class *Filicineae*

Adder's tongue

Ophioglossum vulgatum

SIZE: 5 to 15 inches

An unlikely looking fern, adder's tongue has but a single leaf with a spore spike growing out of its center. It is commonly found in wet soil.

20

Ostrich fern

Matteuccia struthiopteris

SIZE: Up to 5 feet

This fern is usually seen in marshes and on pond shores. The leaflets are alternating. The spore-bearing fronds are nearly half the height of the plant.

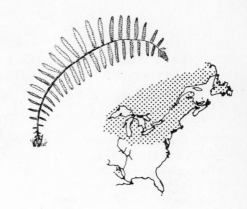

Marsh fern

Thelypteris palustris
(Dryopteris sp.)

SIZE: 2½ feet

The marsh fern spreads rapidly and crowds out other plants, making it unsatisfactory for a garden. The leaflets are opposing at the base of the frond, becoming alternating toward the top. It dies with the first killing frost.

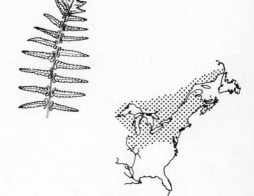

Fancy fern

Dryopteris intermedia

SIZE: 20 inches

This fern is suitable for the garden. The fronds are opposing. It is found in high areas in marshes and swamps.

21

Eastern chain fern,
Virginia chain fern

Woodwardia virginica

SIZE: Up to 4 feet

W. virginica prefers very wet
soils and mud. The leaflets are
alternating on the frond. Spore
cases grow in double rows on
the undersides of the leaflets.

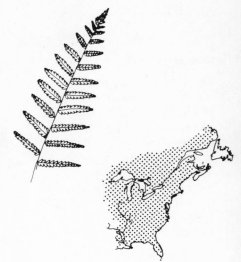

FLOWERING PLANTS—Class *Angiospermae*

Common cattail,
broadleaf cattail

Typha latifolia

SIZE: 3 to 7 feet

This is a familiar marsh
cattail. It can be distinguished
from other cattails by its
rough, bare gray stem
protruding from the spike. The
spike is green-brown to
red-brown when mature.

Narrowleaf cattail

Typha angustifolia

SIZE: 4 to 5 feet

As the name implies, this species differs from *T. latifolia* in having very slim leaves, not much wider than ¼ inch. The spike, shorter than the surrounding leaves, is dark brown when mature (female plants).

Burreed

Sparganium sp.

SIZE: Up to 4 feet

The burreed lives in wet soil or shallow water. The leaves are narrow, averaging an inch in width. The flower clusters become green or brown seed balls. Individual seeds are needed to identify the various species.

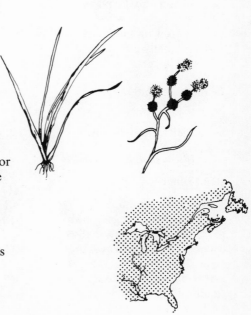

Seaside arrowgrass

Triglochin maritima

SIZE: 3 inches to 3 feet

Seaside arrowgrass is found in both fresh and salt marshes everywhere except in the south. Small green flowers produce seed pods in the fall.

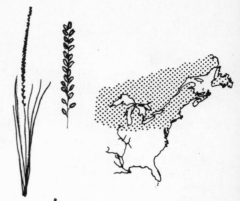

Narrowleaf waterplantain

Alisma geyeri

under water   out of water

SIZE: Up to 2 feet

The spear-shaped leaves are about 4 inches long on terrestrial plants but up to 3 feet long on aquatic plants. The small rosy or white flowers grow on bushy stalks. It occasionally inhabits brackish waters.

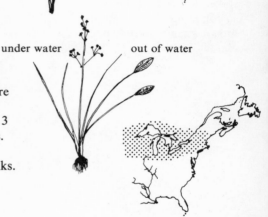

Broadleaf waterplantain

Alisma plantago-aquatica

SIZE: 3 inches to 3 feet

This species, common throughout our range, grows in very moist soil and shallow water. The oblong leaves are generally rounded or heart-shaped at the base. The small white flowers grow on bushy stalks.

24

Creeping burhead

Echinodorus cordifolius

SIZE: 6 inches to 2 feet

E. cordifolius grows on pond shores and in any damp soil. The leaves are arrowhead-shaped and rounded at the base. Flower clusters appear on a tall stalk or (*var. lanceolatus*) a creeping stalk.

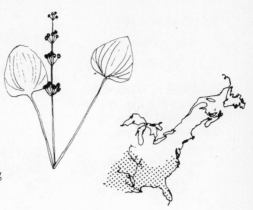

Bur arrowhead

Sagittaria rigida

SIZE: 2 inches to 2 feet

Bur arrowhead leaves vary considerably; they may be ovaloid, lanceolate, or arrowhead-shaped. Small flowers and seed clusters grow on a flexible stem.

Bulltongue

Sagittaria angustifolia

SIZE: 2 to 7 feet

Bulltongue, a southern plant, rarely ranges north of the Chesapeake. It inhabits freshwater and slightly brackish water marshes. The leaves are ovaloid with pointed tips.

25

Broadleaf arrowhead

Sagittaria latifolia

SIZE: 1 to 6 feet

S. latifolia is one of our commonest marsh plants. It has arrowhead-shaped leaves that may be as long as 20 inches. The small white flowers and seed balls grow on a stalk.

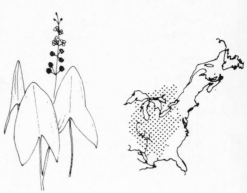

Skunk cabbage

Symplocarpus foetidus

SIZE: Leaves up to 30 inches

Both the common and the Latin names of this plant point out its most obvious characteristic, its skunklike odor. The heavy leaves grow in spiral fashion around the fruit.

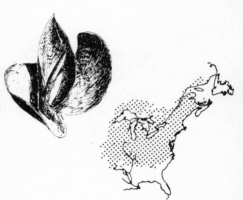

Flowering rush

Butomus umbellatus

SIZE: 3 to 5 feet

This is a rather scarce but attractive import from Europe. Pink flowers, about an inch across, cluster at the top of the stem. Underwater plants do not develop flowers.

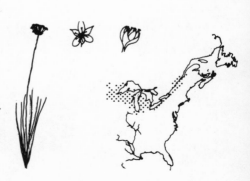

26

American frogbit (frog's-bit)

Limnobium spongia

SIZE: Leaves up to 3 inches

There are both aquatic and terrestrial forms of this plant. The aquatic forms prefer stagnant water; they produce soft, spongy, heart-shaped leaves and are proliferous by runners. Terrestrial plants have upright stems and ovaloid leaves with pointed tips.

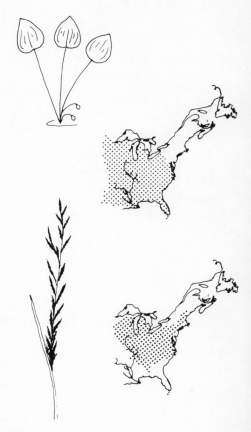

Sharpscale mannagrass

Glyceria acutiflora

SIZE: 1 to 4 feet

The long flower clusters consist of six to twelve flowers with sharply pointed scales.

Phragmites, reed

Phragmites communis

SIZE: 6 to 18 feet

One of our largest and most conspicuous wetland plants, it inhabits both freshwater and brackish water marshes. The 18-inch flower clusters have a blue-to-purple tinge. As they age, they become white and cottony.

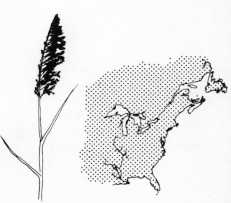

27

Prairie cordgrass,
common cordgrass,
freshwater cordgrass

Spartina pectinata

SIZE: 3 to 6 feet

This is a grass seen often in
both freshwater and brackish
water marshes. It is harvested
for feed and bedding. Dead
plants bend over and form a
thick mat.

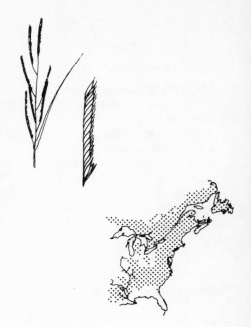

Rice cutgrass

Leersia oryzoides

SIZE: 2 to 5 feet

This rice is common in the
eastern part of our range.
Leaves and sheaths are
extremely rough. The leaves
are as long as a foot; the
flower clusters, up to 8 inches.

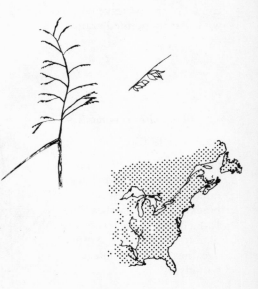

Rice, wildrice

   Oryza sativa

SIZE: 3 to 6 feet

   Zizania aquatica

SIZE: 2 to 12 feet

O. sativa is cultivated in
marshes. It has leaves of up to
18 inches, and flower clusters
nearly as long. The flower
spikelets are rough to the
touch.

Z. aquatica is sometimes
planted as food for wildlife.
The leaves are very long and
narrow, with rough edges. The
seed cluster is comprised of
stiff, upright upper spikelets
(female) and bent-over lower
spikelets (male).

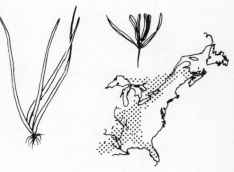

Giant cutgrass

   Zizaniopsis miliacea

SIZE: 3 to 12 feet

The large flower clusters are
widely spaced and droop from
the stem. The leaves are coarse
to the touch.

Jointgrass

Paspalum distichum

SIZE: Branches 3 inches to 2 feet

Jointgrass grows in both freshwater and brackish water marshes. The stems develop horizontally and the branches are upright, supporting double-spiked flower clusters.

Maidencane

Panicum hemitomon

SIZE: 2 to 6 feet

Maidencane often grows in water as much as a foot deep. The flower clusters are up to a foot long. It is prevalent in the southern half of our range.

Sacciolepis

Sacciolepis striata

SIZE: 1 to 6 feet

This is a southern marsh plant. It has long flower clusters that drop early.

30

Wild millet

Echinochloa crusgalli

SIZE: 2 to 6 feet

Wild millet prefers the drier marshes. It has leaves of up to 18 inches and long flower stalks with green-to-purple blossoms.

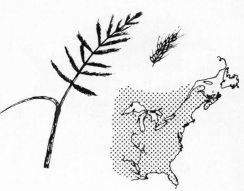

Water spikerush

Eleocharis elongata

SIZE: 3 feet

This is an aquatic southern rush, common in marshes and ponds. Water spikerush has green-brown spikelets.

Walking spikerush

Eleocharis rostellata

SIZE: 3 inches to 4 feet

Those stems growing horizontally reroot at their tips. The spikelets are yellow-brown, becoming green-brown with age.

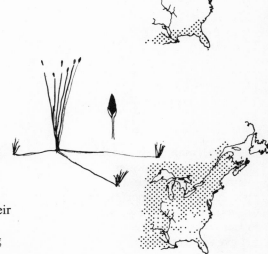

Slender spikerush

Eleocharis acicularis

SIZE: 3 inches to 1 foot

Underwater plants grow longer. Terrestrial plants have clusters of red-and-white-streaked seed pods.

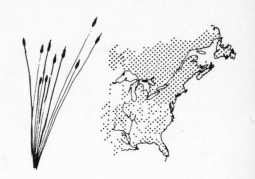

Blunt spikerush

Eleocharis obtusa

SIZE: 3 inches to 3 feet

The blunt spikerush produces single brown spikelets. The plants grow in distinct clumps. This species is found throughout eastern United States and Canada.

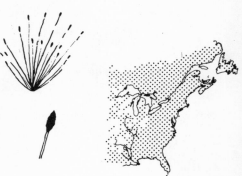

Bluntscale bulrush

Scirpus smithii

SIZE: 3 inches to 3 feet

This rush inhabits all but the southernmost freshwater marshes in our range. The triangular stem supports near its top a cluster of green-to-brown spikelets, above which it bends conspicuously.

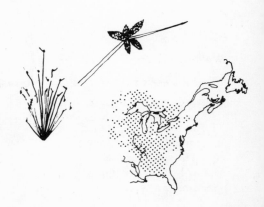

32

Common threesquare

Scirpus americanus

SIZE: 3 to 6 feet

Threesquare is seen in both freshwater and brackish water marshes. Its tall stems are distinguished by their sharply triangular cross section and by the red-brown cluster of spikelets a few inches from the top.

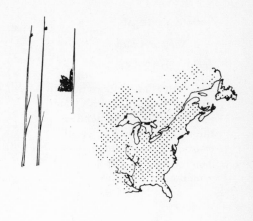

Southern bulrush

Scirpus californicus

SIZE: 6 to 11 feet

Southern bulrush is common to both freshwater and brackish water marshes in the warmer regions. The stems are triangular in cross section; the spikelets are red-brown.

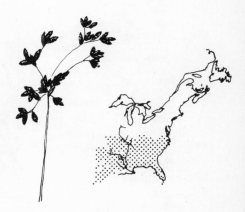

Hardstem bulrush

Scirpus acutus

SIZE: 6 to 12 feet

The hardstem bulrush inhabits nearly any kind of marsh, inland and coastal, in all but the southeasternmost states. The stems are firm and round in cross section. The spikelets are a faded brown.

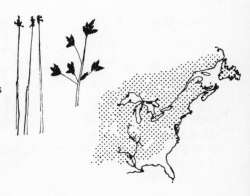

33

Softstem bulrush

Scirpus validus

SIZE: 3 to 12 feet

This bulrush flourishes in both
freshwater and brackish water
marshes. The stems bend in
the wind; they can be easily
crushed between the fingers.
The red-brown spikelets have
a tendency to droop.

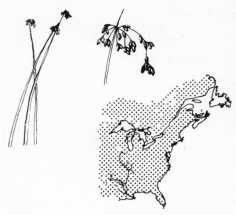

Swamp bulrush

Scirpus etuberculatus

SIZE: 3 to 6 feet

This rush grows primarily in
the inland marshes of the
southern part of our range.
The spikelets are green-brown
when new, becoming dark
brown with age.

Slender bulrush

Scirpus heterochaetus

SIZE: 5 to 10 feet

The slender bulrush is
restricted to freshwater
marshes. The brown spikelet
clusters grow singly instead of
in bunches. It prefers a colder
climate but is sometimes
found south of New Jersey.

River bulrush

Scirpus fluviatilis

SIZE: 3 to 6 feet

The river bulrush is more
frequently encountered on
river and pond shores than in
marshes. The spikelets are
light brown.

Everglade beakrush

Rhynchospora tracyi

SIZE: 2 to 3 feet

This rush prefers the warmer
regions. It produces clusters of
brown spikelets.

Twig rush

Cladium mariscoides

SIZE: 2 to 3 feet

Found in both freshwater and
brackish water marshes, the
twig rush can be recognized by
its long brown flower spikelets.
It commonly grows in sandy
soil.

Lake sedge

Carex riparia

SIZE: 2 to 5 feet

Lake sedge inhabits not only
freshwater marshes but also
pond and lake shores. It
prefers the cooler regions,
rarely ranging south of the
Carolinas.

Slough sedge

Carex trichocarpa

SIZE: 1 to 5 feet

This sedge usually grows in
small clumps in marshes. The
leaves are very rough but not
hairy.

Beaked sedge

Carex rostrata

SIZE: 2 to 3 feet

Beaked sedge prefers shallow
swamp water. The sheaths on
the seed pods show a purple
tinge.

Arrow-arum

Peltandra virginica

SIZE: Up to 3 feet

The leaves of arrow-arum vary
considerably in shape, though
most resemble an arrowhead.
The flower clusters become
pods of red berries on a bent
stem.

Sweetflag, flagroot

Acorus calamus

SIZE: 3 to 6 feet

Sweetflag is most common in
the New England–New York
area. The yellow flower
spikelets grow conspicuously
out of the side of the stem.

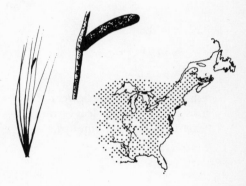

Golden club

Orontium aquaticum

SIZE: 2 feet

Golden club prefers shallow
water. The flower stalk is thick
and nearly white.

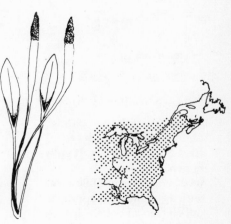

Marsh dayflower

Aneilema keisak

SIZE: 1 to 5 feet

The marsh dayflower grows along the ground or on other plants or objects. Pink blossoms develop at the base of the leaves.

Pickerelweed

Pontederia cordata

SIZE: 2 to 4 feet

Attractive violet-blue flowers grow on a stalk. The leaves are heart-shaped with blunt tips.

Water hyacinth

Eichhornia crassipes

SIZE: Up to 1 foot above water

Water hyacinth is a floating plant that was introduced from South America. It grows in dense colonies. Violet (occasionally white) flowers with an orange dot appear in tall clusters above the leaves.

Roundleaf mudplantain

Heteranthera reniformis

SIZE: Stems 4 to 8 inches

This plant grows in mud or shallow water. The leaves are kidney-shaped. Pale blue or white flowers grow in clusters of three to five, occasionally more.

Soft rush

Juncus effusus var. solutus

SIZE: 3 to 6 feet

The soft rush grows in clumps. Green or brown flowers appear in a fan arrangement partway up the stem.

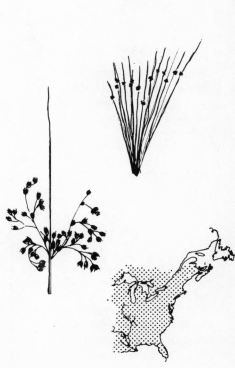

Baltic rush

*Juncus balticus
var. littoralis*

SIZE: 3 to 6 feet

This is a common rush; plants
are evenly spaced rather than
in clumps. The flowers are
green-to-brown.

Creeping rush

Juncus repens

SIZE: Stems 3 to 8 inches

Creeping or floating stems
grow in muddy soil or shallow
water. The green flowers are
less than ¼ inch long.

Soldier rush

Juncus militaris

SIZE: 1 to 3 feet

The brown flower clusters of
this rush grow on a stalk
branching from the main stem.

40

Bog rush

Juncus pelocarpus

SIZE: 3 to 5 inches

This is one of our smaller
rushes. It has clusters of green
flowers, each blossom with
three petals. Underwater
plants are flowerless.

Hairy smartweed

Polygonum hirsutum

SIZE: 2 to 4 feet

Hairy smartweed distinguishes
itself by long hairs on the
stem, as the name implies.
Pink-to-white flowers are
produced in compact masses
on straight spikes.

Marsh smartweed

Polygonum coccineum

SIZE: 2 to 5 feet

Attractive pink flowers grow
on thin, hairy stalks in
marshes. Water-rooted plants
have rounded leaves.

Water smartweed

Polygonum amphibium

SIZE: 2 to 4 feet

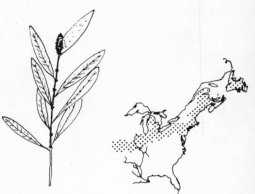

This smartweed may be aquatic or rooted in muddy soil. The flowers are a light red, producing brown seeds. The species adapts to environmental changes such as drying of the soil. It is common in cooler climates.

Southern smartweed

Polygonum densiflorum

SIZE: 2 to 6 feet

Southern smartweed, as the name indicates, ranges in the warmer climates. Blossoms are light pink or white; the seeds are black.

Nodding smartweed

Polygonum lapathifolium

SIZE: 2 to 6 feet

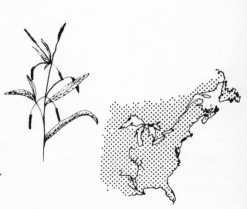

P. lapathifolium varies somewhat in appearance but usually has pink or white flowers on long, curved spikes. The seeds are dark brown or black. It grows in wet soil.

Swamp smartweed

Polygonum hydropiperoides

SIZE: 1 to 4 feet

This smartweed is common in shallow marshes. The white or faintly pink flowers grow on straight stalks; the seeds are black.

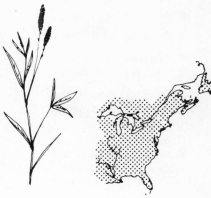

Alligator weed

Alternanthera philoxeroides

SIZE: Stems 18 to 24 inches

Alligator weed is a tropical species with long, oval leaves and white flowers. It grows in muddy soil or shallow water.

Spatterdock, yellow pond lily

Nuphar advena

SIZE: Flower 1 to 3 inches

This water lily grows either in water or in muddy soil. It has leaves of various shape, though they are mostly spadelike. The single yellow flower is greenish on the outside, sometimes showing a violet tinge.

43

American lotus, common lotus

Nelumbo lutea

SIZE: Leaves to 2 feet;
      flowers to 10 inches

The large, variously shaped
leaves that grow along the
water surface are
characteristic of this water lily.
The flowers are light yellow. It
prefers very wet soil or water.

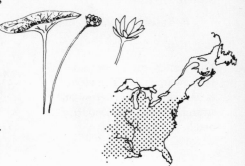

Pitcher plant, huntsman's cup, Indian cup

Sarracenia purpura

SIZE: Up to 24 inches

This unusual plant has
urn-shaped leaves, usually
filled with a watery liquid in
which unwary insects fall and
are digested. The leaves are
suffused with a reddish or
purplish tone.

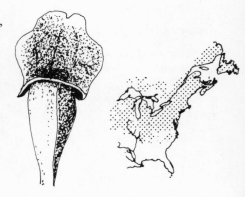

Marsh cinquefoil

Potentilla palustris

SIZE: 2 inches to 2 feet

The leaves, gray-green above
and gray below, have a
sawtoothed border. The
flowers are a strong purple.

44

Marsh hibiscus

Hibiscus moscheutos

SIZE: 2 to 7 feet

Marsh hibiscus is found in
both freshwater and brackish
water wetlands. The leaves are
of variable shape and
rough-edged. The flowers are
white or pink, occasionally
with a red center.

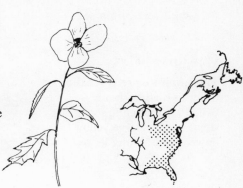

Swamp loosestrife,
water willow

Decodon verticillatus

SIZE: Stems 2 to 7 feet

The leaves are opposing or
arranged in whorls.
Red-purple flowers are
produced at the base of the
leaves. This plant grows in
swamps.

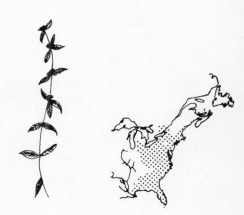

Purple loosestrife

Lythrum salicaria

SIZE: 3 to 6 feet

L. salicaria roots in damp soil,
especially along pond shores.
The red-purple flowers grow
on spikes. The leaves are
lanceolate, sometimes
appearing in whorls of three.

Floating water primrose

Jussiaea repens

SIZE: Up to 10 feet

This primrose prefers shallow water or muddy shores. Creeping or floating root stems support large yellow flowers and oblong leaves.

Marsh purslane, water purslane

Ludwigia palustris var. americana

SIZE: 4 to 6 inches

Marsh purslane is found along shorelines and in other wet soil. The leaves are smooth and oval; red-to-green flowers grow at the base of the leaf stems. Young aquatic plants are reddish.

Parrotfeather, water milfoil, water feather

Myriophyllum brasiliense

SIZE: Leaves 1 to 2 inches

This is a primarily aquatic species that was imported from South America and has become established in the southern part of our range. It has featherlike gray-green leaves. Parrotfeather is a popular aquarium plant.

46

Cutleaf mermaid weed

Proserpinaca pectinata

SIZE: 3 to 18 inches

Mermaid weed prefers sandy
soil, whether in or out of
water. Its leaves are feathery.
The seeds grow at the base of
the leaves.

Marestail

Hippuris vulgaris

SIZE: 3 inches to 3 feet

Marestail is found at the edge
of ponds and streams. The
leaves grow in whorls, with a
minute flower at the base of
each leaf.

Leather leaf, cassandra

Chamaedaphne calyculata

SIZE: Up to 5 feet

In the spring, leather leaf has
clusters of small, bell-shaped
white flowers on the upper
branches. The narrow,
leathery-textured yellowish
leaves are scaly beneath. This
is a predominately northern
plant, occurring in the south
only at higher elevations.

Bulblet water hemlock

Cicuta bulbifera

SIZE: 2 to 5 feet

This species is common in
swamps. Small white flowers
grow on spiny stems. The
roots are extremely poisonous
to animals and man.

Water parsnip

Sium suave (*cicutaefolium*)

SIZE: 3 to 6 feet

This is a common marsh plant
throughout much of our range.
It has sawtooth-edged leaves
and small white flowers which
grow in clusters.

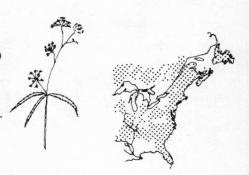

Mock bishopweed

Ptilimnium capillaceum

SIZE: 3 to 18 inches

P. capillaceum is both a
brackish water and a
freshwater plant. It has tiny
white flowers and threadlike
leaves.

Water willow

Justicia americana

SIZE: 2 to 3 feet

Water willow usually grows in clumps in shallow water. It has white flowers with red spots.

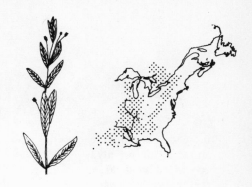

Marsh boltonia

Boltonia asteroides

SIZE: 2 to 7 feet

This plant is found in the moist soil bordering streams and ponds. A pink or white flower grows at the end of each of the many stems.

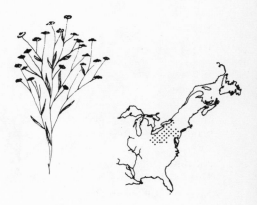

Nodding beggartick

Bidens cernua

SIZE: 3 inches to 6 feet

Small yellow flowers with dark centers grow singly or in clusters. The leaves are large, pointed, and rough-edged. *B. cernua* is common in all but the southernmost parts of our range.

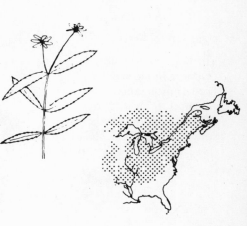

Swamp lily

Saururus cernuus

SIZE: 4 feet

In the southernmost reaches of
our range, swamp lily is
recognized by its large,
fragrant white flowers.

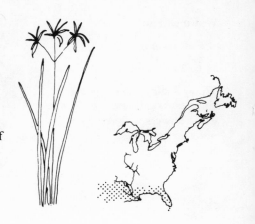

American hornbeam, blue beech, water beech

Carpinus caroliniana

SIZE: Up to 40 feet

The hornbeam is most
common in moist soil
bordering streams and
wetlands. It has catkins like a
birch, sawtoothed leaves, and
seed-bearing leaf clusters
(bracts). The smooth bark is
blue-gray.

Red maple

Acer rubrum

SIZE: 70 to 80 feet

The red maple is the most important contributor to the colorful foliage of autumn. The leaves have three coarse-toothed lobes which turn bright red and yellow as winter approaches. In early summer there are red-to-yellow flower clusters on the twigs. Later in the season the paired seeds, shaped like a moustache, appear. This tree grows best in swampy soil.

CONIFERS—Class *Coniferinae*

Tamarack, larch

Larix laricina

SIZE: 40 to 80 feet

Unlike the common pine, the tamarack is deciduous. In the fall the trees lose their blue-green color, turning yellow before the needles fall. The cones are ovoid, $\frac{1}{2}$ to $\frac{3}{4}$ inch long, with twelve to fifteen scales. The tamarack prefers a moist, boggy soil, though it may be found in nearly any environment.

Atlantic white cedar

Chamaecyparis thyoides

SIZE: 80 feet

White cedar is found in freshwater wetlands near the Atlantic coast. The branchlets have small, overlapping blue-green scales. The cones are blue and wrinkled; the bark, gray to red-brown.

Eastern red cedar

Juniperus virginiana

SIZE: 40 to 50 feet

J. virginiana prefers wet, swampy soil. It is actually a juniper. The bark is gray to red-brown and fibrous. The round or four-sided branchlets are covered by overlapping deep green scales. Cones are ⅓ inch in diameter, turning blue with maturity.

Also known as bullheads, horned pout can survive in very shallow water by burrowing in the mud. Though not truly a marsh animal, a fish may find its way into a marsh during high water.

Aquatic nymph of a dragonfly (*Anax sp.*). This specimen measures 1¼ inches.

The crayfish is found over most of the United States. It is a freshwater crustacean that prefers a temperate climate.

The painted turtle is usually found in ponds, ditches, quiet brooks, or bogs. It leaves the water only for migration and rest. (Scott-Swedberg)

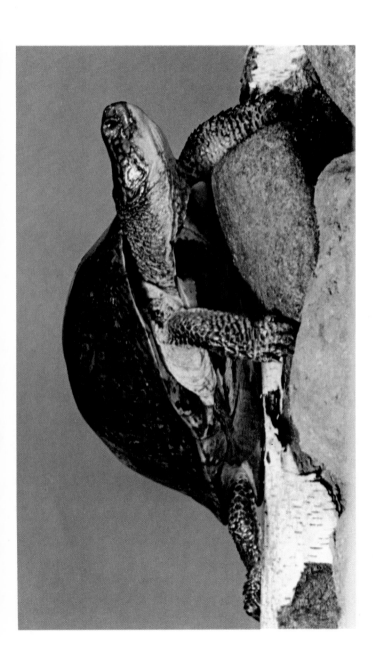

The Blanding's turtle prefers to live in or near water. It may be
found, however, at a considerable distance from any body of water.

The great blue heron is a patient hunter. It may wait half an hour or
more, still as a statue, until a meal comes within reach. (Scott-Swedberg)

The white-tailed deer frequently wades to graze on
submerged aquatic vegetation. (John E. Swedberg)

Canada geese may fly as far as 4,000 miles during
their annual migration. (John E. Swedberg)

The moose is the largest visitor to the wetlands.
It is seen only in northerly areas. (John E. Swedberg)

The spotted salamander is rarely seen out in the open in daytime.
It prefers to hide beneath leaves and sticks near a pond or stream.

The dragonfly may well become the next meal for the young flycatchers, the bullfrog, or the tree swallow. The abundance of aquatic insects in the marshes is one of the reasons for the great number and variety of animals in this environment. (John E. Swedberg)

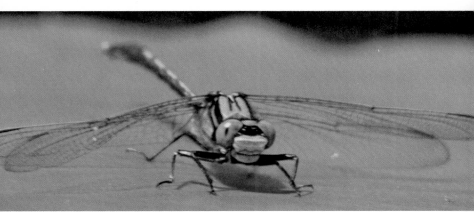

Newly hatched fish may take refuge in
shallow marsh water to escape predators.

The beaver probably has more influence on the wetlands than any
other animal: it literally changes their form by building its familiar dams.

The damselfly spends the first stage of its life as an aquatic nymph. After a
year or more it crawls out of the water onto a plant or rock and sheds its
nymphal skin, becomes somewhat larger, and develops its color, all in a
matter of minutes. (All photographs are at the same magnification.)

ANIMAL LIFE IN THE FRESHWATER WETLANDS

Kingdom ANIMALIA—ANIMALS

Phylum PROTOZOA—Plankton
    Class *Mastigophora*—Flagellates
Phylum PORIFERA—Sponges
Phylum COELENTERATA—Hydras and jellyfishes
Phylum BRYOZOA—Moss animals
Phylum PLATYHELMINTHES—Flatworms
Phylum ANNELIDA—Segmented worms
Phylum MOLLUSCA—Snails and mussels
Phylum ARTHROPODA—Arthropods
    Class *Insecta*—Insects
    Class *Crustacea*—Crustaceans
    Class *Arachnida*—Spiders
Phylum CHORDATA—Chordates
  Subphylum VERTEBRATA—Vertebrates
    Class *Amphibia*—Amphibians
    Class *Reptilia*—Reptiles
    Class *Aves*—Birds
    Class *Mammalia*—Mammals

There is as good a representation of the animal world in freshwater wetlands as in most other environments. Specimens range in size from microscopic to the mammoth proportions of the moose. Some wetland species live in the water, some near it, others in saturated soil.

The single-celled organisms (plankton) are discussed in the first part of the book.

SPONGES—Phylum PORIFERA

Spongilla sp.

Sponges are colonial animals.
Freshwater sponges hardly
resemble their saltwater cousins.
They are small, jellylike masses
found attached to sticks or rocks.
Most species die in cold weather,
leaving buds (gemmules) to start
new colonies the next year.
Sponges do not need sunlight for
growth and therefore are often
found under objects. Lacking
internal organs, they feed by
filtering microscopic organisms
from the water as it flows
through them.

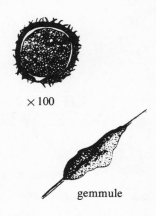

× 100

gemmule

HYDRAS, JELLYFISHES—Phylum COELENTERATA

Hydra sp.

SIZE: Up to 1 inch

Hydras are freshwater cousins
of the more familiar sea anemones,
marine jellyfishes, and corals. They
feed by capturing and stinging
their prey with their tentacles.

Freshwater jellyfish

Craspedacusta sowerbyi

SIZE: ½ inch

This is a somewhat rare animal
in North America, and *C. sowerbyi*
is the only species considered
natural to our range. When young
it looks much like a hydra.

54

MOSS ANIMALS—Phylum BRYOZOA

Pectinatella magnifica

SIZE: ¼ inch (zooid)

Moss animals are rare in
polluted waters, but where
water is clean they are prolific
to the point that they may clog
water pipes. They grow in
encrusting colonies and feed
on microscopic organisms.

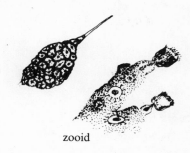

zooid

FLATWORMS—Phylum PLATYHELMINTHES

Planaria

Dugesia dorotocephala

SIZE: 1 inch

This is the flatworm most
often found in the wetlands. It
is free-living rather than an
attached parasite. *D.
dorotocephala* is the common
flatworm studied in biology
classrooms. Trying to identify
worms can be a challenge, as
some are easily confused with
certain insect larvae.

SEGMENTED WORMS—Phylum ANNELIDA

Leech, bloodsucker

Macrobdella sp.

SIZE: Up to 10 inches
(sometimes more)

The leech, or bloodsucker,
lives on the blood of aquatic

55

vertebrates. It is distinguished by a row of black-and-red spots on the back.

Bristleworm

Aeolosoma sp.

SIZE: Up to ½ inch

Chaetogaster sp.

SIZE: Up to ½ inch

Aeolosoma sp. is an aquatic earthworm. It is commonly called bristleworm because of the bundles of bristles (setae) that grow on each of its segments.

Chaetogaster sp. often lives on or in mollusk shells. Most specimens are colorless and nearly transparent.

SNAILS, MUSSELS—Phylum MOLLUSCA

Little pond snail

Amnicola limnosa

SIZE: ⅓ inch

The little pond snail is found in pools, ponds, and streams. It is a wide-ranging species and may be discovered almost anywhere east of the Rockies. The color varies from tan to deep brown.

Winkle

Viviparus intertextus

SIZE: 1½ inches

This is a dark-colored snail
that prefers muddy bottoms.
Several species of *Viviparus* are
found in eastern North
America.

Hairy wheel snail

Gyraulus sp.

SIZE: ⅓ inch

A flat, wheellike shell and a
coating resembling hair
distinguish this snail.

Giant pond snail

Lymnaea stagnalis

SIZE: 2½ inches

This is a large snail that
prefers quiet water. It is found
in central North America. The
shell is thin and fragile.

Papershell

Anodonta grandis

SIZE: Up to 6 inches

The papershell is generally
distributed throughout our
range. It has a thin greenish
shell.

Pearly mussel

Elliptio crassidens

SIZE: Up to 6 inches

This is a heavy, rough-shelled
clam common in eastern
North America. The valves
can be cut and polished to
make pearl ornaments and
buttons.

ARTHROPODS

Arthropods **(Phylum ARTHROPODA)** constitute the largest of all
phyla. They include the insects (more than 900,000 species), the
crustaceans, and the spiders and their relatives.

The life cycles of these creatures are complex, particularly the
aquatic species, which may go through a number of stages before
emerging as adult. Some are vegetarian, some carnivorous; others
have specialized feeding habits, such as blood-sucking mosquitoes.
Arthropods in turn are a food source for numerous other animals.

Insects have played a major role in the spread of communicable
diseases. They also can raise havoc with crops. Yet their presence is
necessary if our songbirds and fish are to sustain life.

Crustaceans are largely marine but are well represented in freshwater
wetlands too. They include the crabs and crayfish as well as scorpions,
sowbugs, water fleas, and shrimps.

Spiders are a subject of superstition and fear, yet only a very small
percentage of them are harmful to man. There are actually no true
aquatic spiders, but some species live in close association with a
wetland environment.

INSECTS—Class *Insecta*

Giant water bug

Lethocerus americanus

SIZE: Up to 2½ inches

This large aquatic insect can inflict a painful bite on a swimmer. It has been known to invade fish hatcheries and cause serious losses. Color varies from dark brown-green to brown-yellow, with a velvety silver underside. Though strictly a nocturnal flyer, *L. americanus* is readily attracted to light and is sometimes found stranded at the base of light poles near wetlands.

Toad bug

Gelastocoris oculatus

SIZE: Slightly over ¼ inch

This small yellow-brown insect resembles a tiny toad both in shape and in its tendency to hop. It is found along the water's edge, where it feeds on smaller insects. It lays its eggs in sand or mud. The toad bug ranges from New England southward.

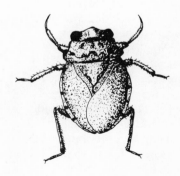

Water boatman

Arctocorixa alternata

SIZE: ½ inch

As the name implies, the water boatman is an excellent swimmer. At the first sign of danger it dives to the bottom, and can often be seen clinging to submerged plants. *A. alternata* is a mottled gray-and-black. The adult is a nocturnal flyer. The larvae go through five changes (instars) before molting into adults. Most commonly found in the northern part of our range, the water boatman is vegetarian.

Backswimmer

Notonecta sp.

SIZE: ½ inch

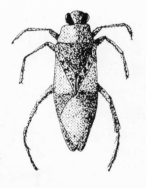

The backswimmer is unique among the aquatic insects in that it swims upside down. An active nocturnal flyer, it is predatory, living on small aquatic creatures, and is capable of inflicting a painful bite if handled. *Notonecta sp.* is brownish-black and white with ruby eyes. Eggs are attached to submerged plants and hatch in two weeks. The larva passes through five instars before emerging as an adult.

Water scorpion

Ranatra sp.

SIZE: 1 inch

The water scorpion is a poor
swimmer and would rather
walk about on aquatic
vegetation. It can be
recognized by its long legs
with claws, sticklike body, and
taillike breathing organ. It has
wings but rarely takes flight,
and then only at night.
Ranatra sp. is predatory,
feeding on nymphs, mosquito
larvae, and small crustaceans.
Eggs, deposited in dead or
living plant tissue, hatch in
two to four weeks. The larvae
pass through five instar stages
to become adults in about five
weeks.

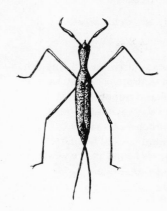

Water strider, water skater,
pond skater

Gerris sp.

SIZE: ½ to ¾ inch

Throughout most of North
America, the water strider
rapidly skitters across lakes,
ponds, pools, and slow
streams. It is a dark, slender
insect. The adult is usually
wingless. *Gerris sp.* is
nocturnal and preys on other
aquatic insects.

Marsh treader

Hydrometra martini

SIZE: Up to ½ inch

The marsh treader is a sluggish insect that preys on aquatic larvae and pupae. It is dark brown-to-black, and looks very much like a small animated stick. The marsh treader is found in the western part of our range.

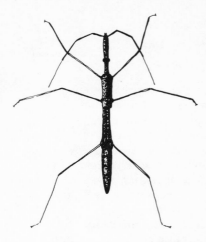

Mayfly, shadfly

Hexagenia sp.

SIZE: ½ to 1 inch

The attractive adult mayfly lives for only a span of hours, at most a few days, during which it does not eat. Adults mate and deposit eggs in the water, which sink to the bottom. The eggs hatch into aggressive larvae, which may remain active for a year or more. Both the larval and adult stages are an important food for many other animals.

larva

Dragonfly, darner

Anisoptera

SIZE: 1½ to 4 inches

The dragonfly is readily distinguished by its long, slim body and broad, symmetrical wings which remain extended when at rest. It is a strong flyer and easily catches its prey in midair. The prominent compound eyes appear to be too large for the head. The dragonfly is a beneficial insect, preying upon such noxious pests as black flies, gnats, and mosquitoes. The immature nymphs (naiads), which are short and heavy-bodied, go through many molts in the water before emerging as adults. The nymphs are aggressive aquatic predators.

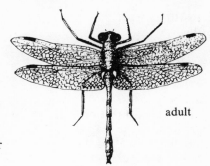

adult

Anax sp.

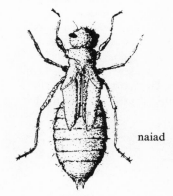

naiad

Libellula sp.

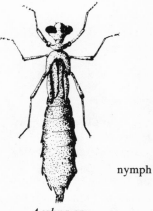

nymph

Aeshna sp.

Damselfly, darner

Zygoptera

SIZE: 1 to 2 inches

The damselfly is a familiar insect common to nearly every body of water. It resembles its larger relative, the dragonfly, except that its body is slimmer and its eyes larger in proportion. It has brilliantly colored, often metallic-looking markings on the body and somewhat club-shaped wings. Unlike the dragonfly, the damselfly generally folds its wings back when at rest. The naiad is characterized by three featherlike tails (tracheal gills).

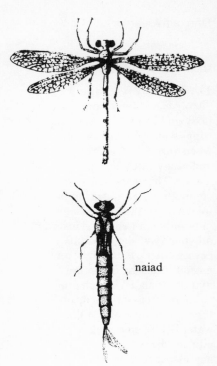

naiad

Alderfly

Sialis sp.

SIZE: ½ inch

This is a dark brown-to-gray insect common around wetlands. The larvae are usually found in moving water; they crawl out to pupate in wet soil. The adult lives only a few days.

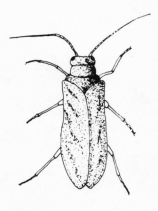

Dobsonfly, hellgrammite

Corydalus cornutus

SIZE: 4 to 5 inches

This conspicuous insect is
probably best known in its
larval stage, the hellgrammite,
a popular fishing bait. The
adult, one of the largest
aquatic insects, is dark gray. It
is a poor flyer. The eggs are
laid on rocks and branches at
the water's edge. The larvae
prefer moving water, where
they may remain two or three
years before emerging as
adults.

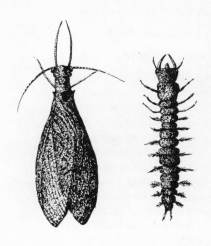

Caddisfly

Trichoptera

SIZE: Up to 1 inch

The caddisfly is usually
hairy-winged and
gray-to-brown. When at rest it
characteristically folds its
wings over its body in rooflike
fashion. Caddisflies are
generally nocturnal. The
larvae of most species build
cases of sticks, pebbles, or
vegetative debris by secreting
an adhesive and silk threads.
They remain mostly in the
case, pulling themselves about
by their front legs and
withdrawing at the first sign of
danger.

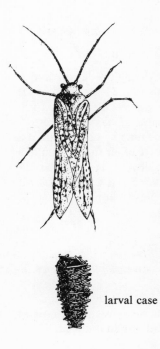

larval case

65

False longhorn leaf beetle

Donacia piscatrix

SIZE: ¼ inch

In eastern North America, *D. piscatrix* lives in close association with flowering aquatic plants, water lilies in particular. The eggs are laid beneath the surface of the water and hatch into grublike larvae. Each larva spins a cocoon on a submerged plant and hatches as an adult in about ten months. The leaf beetle feeds on aquatic vegetation.

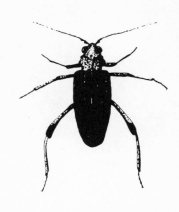

Diving beetle

Dytiscus fasciventris

SIZE: 1 inch

The diving beetle will dive beneath the water surface at the slightest disturbance. It is capable of staying under for some time by taking with it a reserve of air in the form of a bubble attached to its tail. *D. fasciventris* is green-black with a yellow margin.

Rice water beetle

Lissorhoptrus oryzophilus

SIZE: 1 to 1⅛ inches

This beetle is aquatic both as an adult and in the larval

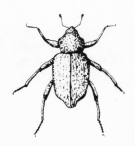

stage, when it is sometimes
known as the rice-root
maggot. The adult is
gray-brown with a coating of
scales. The larva is white and
legless. It can be a serious pest
in rice fields.

Whirligig beetle

Dineutus ciliatus

SIZE: ½ inch

This active insect never seems
to run out of energy, scooting
around on the surface of the
water. It will dive if it can't
outrun danger, and it can fly
at will. It is a shiny, metallic
black.

Striped diving beetle

Agabus disintegratus

SIZE: ⅓ inch

The striped diving beetle is
easily identified by its
brown-yellow color with two
black crossbands behind the
head and black lengthwise
stripes on the wing covers. It is
predatory and readily attacks
small aquatic animals,
including polliwogs.

Water scavenger beetle

Berosus striatus

SIZE: ⅛ to ¼ inch

This beetle has a green-yellow tinge with elongate small black spots. *B. striatus* larvae are vicious predators and often cannibalistic.

Minute water scavenger beetle

Hydrochus scabratus

SIZE: ¼ inch

Hydrochus squamifer

SIZE: ⅛ inch

These are small beetles with a metallic bronze sheen. They are found along the edges of freshwater or brackish water ponds, stagnant pools, and slow streams, feeding on decayed vegetation.

Fasciated diving beetle

Laccophilus fasciatus

SIZE: ¼ inch

This small diving beetle is found in freshwater environments throughout our range. It is green-yellow with a black stripe across the back.

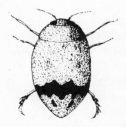

Mosquito

SIZE: 1¼ to 1¾ inches

Throughout history the mosquito has influenced man. Not only has this insect been a pest but, more important, some species are carriers of such fatal diseases as yellow fever, malaria, filariasis, and encephalitis. Mosquitoes are generally associated with wetlands, marshes and bogs in particular. These are their prime breeding grounds, though any standing water, even if miles away from a marsh, is suitable habitat for the propagation of many mosquito species. Eggs laid in the water hatch into larvae (wrigglers). The larvae molt into pupae, which in turn shed their aquatic skins to become adults. Mosquitoes differ from other members of their order in having wing scales and in the well-known spikelike proboscis, capable of penetrating animal skin.

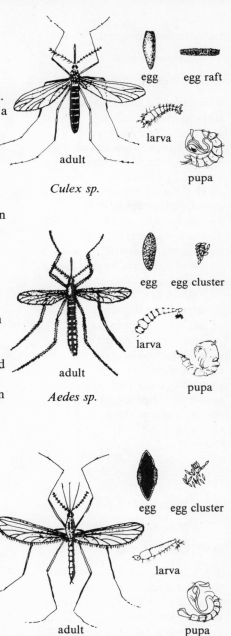

egg   egg raft

larva

adult   pupa

Culex sp.

egg   egg cluster

larva

adult   pupa

Aedes sp.

egg   egg cluster

larva

adult   pupa

Anopheles sp.

69

Black fly, buffalo gnat

Simulium sp.

SIZE: ⅛ inch

What would a campout be
without black flies?
Undoubtedly a bit more
pleasant. Nevertheless, this is
a well-entrenched resident of
the woods and wetlands. The
black fly is not a true marsh
insect. Rather the immature
form is found in small streams.
The larvae attach themselves
to some stationary object and
feed on food particles that are
carried along by the current.
The adults are of medical and
economic importance because
of their disease-carrying
potential, particularly to
domestic animals. Vicious
bloodsuckers, they are
generally present in
marshlands because of the
dense animal population there.

Eastern crane fly

Tipula abdominalis

SIZE: 1 inch

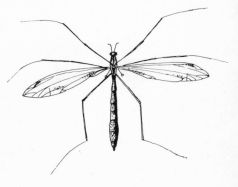

The crane fly is often mistaken
for a giant mosquito, but it is
in fact a harmless vegetarian
insect. This slow-flying
creature is associated with
damp, marshy areas. The

larvae develop in shallow water or muddy soil and decaying vegetation. The adult varies from grayish brown to yellowish brown. At close inspection one can see behind each wing a small projection with a spherical structure at the end.

larva

CRUSTACEANS—Class *Crustacea*

Daphnia, water flea

Daphnia longispina

SIZE: ¹⁄₁₆ inch

D. longispina is probably the best known water flea. In poorly oxygenated pools it appears reddish. Its food consists of algae, microscopic animals, and organic debris (detritus).

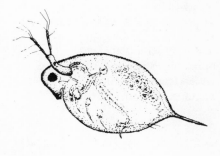

Scud, sideswimmer

Gammarus sp.

SIZE: ⅓ to ½ inch

The scud is a bottom-dwelling invertebrate shaped somewhat like a flea. It is a scavenger and in turn serves as food for larger invertebrates.

Sowbug

Asellus sp.

SIZE: ¾ inch

This is a common bottom-dwelling crustacean that lives on decayed plant matter.

Cyclops

Copepoda

SIZE: ¹⁄₁₀ inch

This tiny semi-planktonic animal lives in open fresh water, feeding on algae, bacteria, and organic debris.

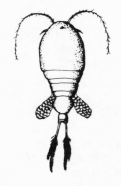

Seed shrimp

Cypridiopsis sp.

SIZE: ¹⁄₁₀ inch

This is a minute crustacean that is very common throughout our range. Under magnification you can see a hairy, bean-shaped creature. It is an important food source for many aquatic animals.

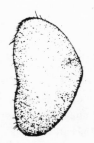

Crayfish, crawfish, crawdad

Cambarus sp.

SIZE: 3 inches

There are upwards of 200 species of crayfishes in North America. A close cousin of the marine lobster, the crayfish is typically aquatic, though many species are able to survive in wet soil.

SPIDERS—Class *Arachnida*

Water mite

Limnochares sp.

SIZE: ¼ inch

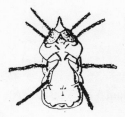

Water mites are semi-planktonic organisms, feeding on smaller invertebrates. *Limnochares sp.* is red. A poor swimmer, it is found in stagnant pools in bogs and swamps.

ventral view

Common water mite, red water mite

Eylais sp.

SIZE: ¹⁄₁₀ inch

Eylais sp. is round and red, with hairy legs. It swims actively.

ventral view

Fisher spider

Dolomedes triton

SIZE: ¾ inch

Spiders are basically
nonaquatic. The fisher spider,
however, earns its livelihood
preying on aquatic insects and
occasionally on tiny fish. It is
capable of diving beneath the
surface and remaining
submerged for considerable
periods.

CHORDATES

Vertebrates

(Subphylum VERTEBRATA)

The vertebrates, animals with a backbone, are the largest and most
conspicuous members of the wetland community. Their presence is
made known by their tracks, nests, dams, and sounds, as well as by
their physical size.

AMPHIBIANS—Class *Amphibia*

Hellbender

Cryptobranchus alleganiensis

SIZE: 18 inches

The hellbender is a strictly
aquatic amphibian. It is easily

distinguished by its gills and
wrinkled, loose-fitting skin.
The hellbender is most usually
found in the fresh waters of
the Ohio Valley.

Mudpuppy

Necturus maculosus

SIZE: Up to 16 inches

This is a salamander. It has
three pairs of red external gills
and four legs. The mudpuppy
lives in nearly any body of
water. It does not hibernate.

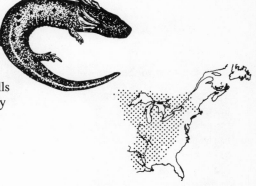

Redbacked salamander

Plethodon cinereus

SIZE: 4½ inches

P. cinereus is a slim, shiny
salamander that is terrestrial;
it is found under moist leaves,
logs, and the like. It has two
color phases: one with a
reddish line down the back,
the other with a grayish line.

Two-lined salamander

Eurycea bislineata

SIZE: 3 to 4 inches

This salamander, streamlined
and glistening, is recognized
by two dark lines running
down the sides of the back.
The center of the back varies
from yellow to brown-red. It is
a rapid swimmer.

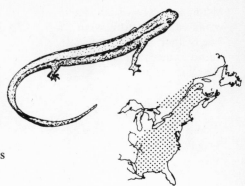

Spotted salamander

Ambystoma maculatum

SIZE: Up to 8 inches

The spotted salamander
occurs everywhere in our
range except the extreme
north and subtropical south. It
lives in soft, wet soil or damp
mulch, and breeds in ponds
and slow streams, laying large
masses of eggs. The bright
yellow spots distinguish it
from other salamanders.

Eastern newt

Notophthalmus viridescens

SIZE: Up to 4½ inches

This amphibian breeds in the
water. The aquatic immature
form has external gills. It
develops into an eft and takes
to terrestrial life for two to
three years. It then returns to

the water and remains aquatic.
The newt is yellow-green with
red spots in the water;
red-orange with red spots on
land (eft stage).

Common treefrog,
gray treefrog

Hyla versicolor

SIZE: Up to 2½ inches

The treefrog has rough, warty
skin and webbed hind feet.
The skin can change color
from gray-brown to
gray-green. It has distinctive
orange markings under the
hindparts. *H. versicolor* is not
restricted to the wetlands; it
may be found in nearly any
environment.

Chorus frog, swamp chorus frog,
swamp cricket frog

Pseudacris nigrita

SIZE: Up to 2 inches

The chorus frog favors a
moderate climate except in the
western part of our range,
where its area extends into
Canada. Its voice resembles
that of the spring peeper but is
more trilled. This frog varies
from green-gray to brown
above, with three irregular dark
stripes running down the back.

Spring peeper

Hyla crucifer

SIZE: Up to 1¼ inches

The loud call of the spring peeper is a well known sound of the season. Most of these little frogs show a distinctive "X" on their backs. *H. crucifer* has a pointed snout and varies from orange-tan to tan.

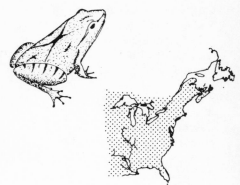

Pickerel frog

Rana palustris

SIZE: Up to 3 inches

Rows of closely spaced, squarish dark markings on the back and an orange tone on the undersides of the hind legs clearly identify the pickerel frog. It is poisonous for many reptiles and other amphibians.

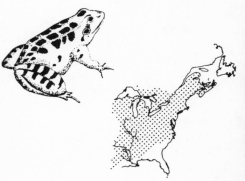

Leopard frog

Rana pipiens

SIZE: Up to 4 inches

R. pipiens is the most common frog in our range. It prefers swampy areas but may be found nearly anywhere. The color varies somewhat, but round or oval spots above are an identifier. The hind legs have dark bands; the underside is whitish.

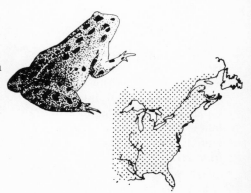

Bullfrog

Rana catesbeiana

SIZE: Up to 8 inches

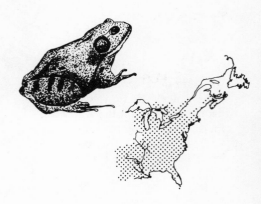

The bullfrog is North America's largest frog. It has large ear discs and a green-brown back that becomes greener toward the head. There are dark crossbands on the hind legs. Bullfrog tadpoles take two years to mature.

Green frog

Rana clamitans

SIZE: Up to 4 inches

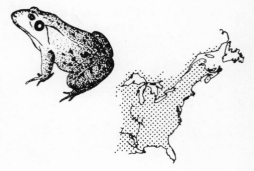

The green frog looks like a small bullfrog except that it is slimmer. It usually has spots on the lower abdomen. *R. clamitans* rarely leaves the water for any extended time.

American toad

Bufo americanus

SIZE: Up to 4 inches

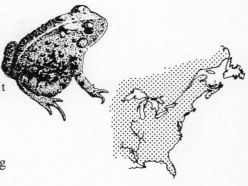

This is our commonest toad. It varies considerably in color, but in general the skin is yellowish, with spots underneath, and has large warts. In the spring it lays long strings of eggs in the water.

Fowler's toad

Bufo fowleri

SIZE: Up to 4½ inches

The skin of this toad is dry in appearance. The Fowler's is distinguished from the American toad in having more and smaller warts as well as a distinctive white line down the back.

Eastern spadefoot

Scaphiopus holbrooki

SIZE: Up to 3 inches

Spadefoot toads are named for a spade-shaped black digging blade on the forefoot. The eastern spadefoot is brownish to green-yellow and often has two light lines down the back. A burrowing toad, it lives in almost any environment, but takes to water to mate and lay its eggs.

REPTILES—Class *Reptilia*

Painted turtle

Chrysemys picta

SIZE: Up to 8 inches

The painted turtle is found in both freshwater and brackish water wetlands. The smooth carapace with red markings on the edge and the red streaks on the head and neck clearly distinguish it from other turtles.

80

Spotted turtle

Clemmys guttata

SIZE: Up to 5 inches

This turtle is easily recognized by its bright yellow spots. It lives on insects and other small animals. Unfortunately its mild manner makes it a popular pet and hence liable to capture.

Blanding's turtle

Emys blandingi

SIZE: 9 inches

Its paucity and its secretive habits combine to make the Blanding's a rarely observed turtle. It lives in ponds, marshes, and damp forests. It has a black carapace, yellow plastron with black blotches, and yellow throat.

Map turtle

Graptemys geographica

SIZE: Females, up to 10 inches; males, up to 5 inches

The carapace of this species is fairly flat with a slight keel. Faint yellow lines on the shell resemble rivers drawn on a map. The map turtle is secretive and not seen as frequently as most other kinds. It feeds on a variety of aquatic vertebrates.

Musk turtle, stinkpot

Sternotherus odoratus

SIZE: 4 inches

This small dark-colored turtle obtains its name from a fluid with a musky odor which it discharges when disturbed.

Snapping turtle

Chelydra serpentina

SIZE: Carapace to 15 inches

The snapping turtle can weigh as much as 35 pounds, occasionally more. It eats both plants and animals. Although it generally burrows in mud on the bottom of ponds, sometimes it can be observed traveling overland. This species is distinguished by its rough carapace and by a body that seems too large for its shell.

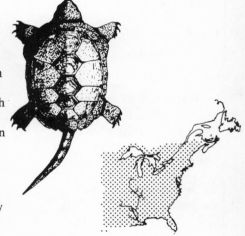

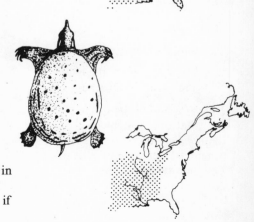

Softshell turtle, smooth softshell turtle

Trionyx muticus

SIZE: 7 inches

The softshell turtle is unique in its pliant, leathery shell. It is predacious, and readily bites if carelessly handled.

Spiny softshell turtle

Trionyx ferox

SIZE: Up to 12 inches, sometimes more

The spiny softshell turtle is recognized by bumps ("spines") on the front of the carapace. It is somewhat larger than *T. muticus*.

Brown water snake

Natrix taxispilota

SIZE: Up to 6 feet

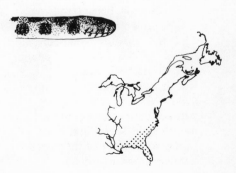

The brown water snake is distinguished by small, bulging eyes, a long, narrow head, and a heavy body. Its color may vary from red-brown to green-brown, with black-bordered dark spots; the belly is whitish with dark markings.

Common water snake

Natrix sipedon

SIZE: Up to 4½ feet

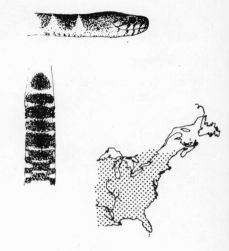

This is our most familiar aquatic snake. It bites viciously if handled. Young snakes have distinctive crossbands across the back, narrow bands on the sides. As the snake matures the overall light brown color becomes darker, until in an old specimen the markings are obscured. *N. sipedon* gives birth to live young.

Cottonmouth, water moccasin

Agkistrodon piscivorus

SIZE: Up to 6 feet

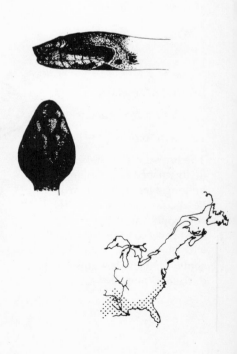

In the southern part of our range, the poisonous cottonmouth stalks turtles, fish, and rodents that wander too close to the water. The adult snake appears black; juveniles show black bands on a dark brown background. The large, diamond-shaped head identifies it as a pit viper and should be adequate warning to the observer to stay away. When disturbed, the cottonmouth often opens its mouth wide, showing the nearly white interior that gives it its common name.

84

Crocodile

Crocodylus acutus

SIZE: Up to 12 feet,
   occasionally more

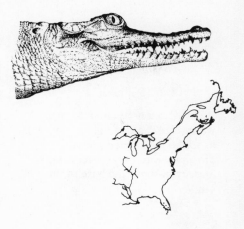

The crocodile has a long, slim
snout; when the mouth is
closed the teeth fit into
pockets outside the jaws. It
readily takes to either salt,
brackish, or fresh water. There
are five toes on the front feet
and four on the webbed hind
feet.

American alligator

Alligator mississipiensis

SIZE: Up to 12 feet,
   occasionally more

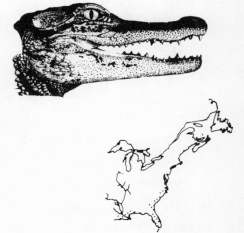

The alligator is distinguished
from the crocodile by its
broader snout and the fact
that its teeth are not visible
when the mouth is closed. All
four feet are webbed. Unlike
the crocodile, the alligator
rarely enters brackish or salt
water.

Pied-billed grebe

Podilymbus podiceps

SIZE: 12 to 15 inches

A short, heavy bill on a brown water bird is a dead giveaway as to the pied-billed grebe. During the spring a black ring appears around the base of the bill.

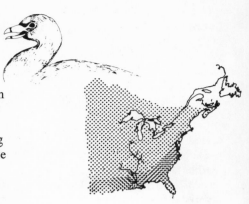

American egret, common egret

Casmerodius albus

SIZE: 40 inches

This bird is considerably larger than any other white heron in our range. It has a yellow bill and black legs. During the breeding season it sports many long plumes on its back.

Great blue heron

Ardea herodias

SIZE: Up to 4 feet

One of the most pleasurable sights for the visitor to a marsh is the great blue heron. This

tall, stately bird has a white
head with black plumes, a
brown-gray neck and black
shoulders, and a blue-gray
body. In flight it can be
distinguished from a crane by
its crooked neck. It feeds
primarily on fish and small
animals.

Green heron

   Butorides virescens

SIZE: 18 to 20 inches

The green heron appears
nearly black, with a red-brown
neck; in bright light its back
shows a deep blue-green. It
has yellow-orange legs.

Black-crowned night heron

   Nycticorax nycticorax

SIZE: 25 inches

This squat, short-legged heron
is a night-feeding bird of the
marshes. It has a green-brown
cap and back, and white head
plumes. An immature bird is
mostly covered with brown
streaks.

American bittern

Botaurus lentiginosus

SIZE: 24 to 35 inches

The American bittern is a large, heavy-bodied bird that prefers to remain on the ground. When threatened it protects itself by remaining motionless, its dark brown back and tan-and-brown-streaked breast blending with the surroundings.

Whistling swan

Olor (Cygnus) columbianus

SIZE: 50 to 58 inches

This large white bird is our most common species of swan. It has a black bill with a yellow spot at its base. "Whistlers" are often seen in company with Canada geese.

Mute swan

Cygnus olor

SIZE: Up to 60 inches

The mute swan is distinguished by its orange bill with a black knob at the base. A European bird introduced here in the past century, it has established itself on the East Coast.

Canada goose

Branta canadensis

SIZE: 22 to 43 inches

A white chin strap is this bird's distinguishing mark. It has a black head and neck, light gray breast, and black tail. This goose may vary considerably in size depending upon its range.

Mallard

Anas platyrhynchos

SIZE: 20 to 28 inches

About the only thing in common between the male and female dress is the violet wing patch bordered with black and white lines. The male mallard has a green head and neck with a white collar, and white tail feathers. The female is generally a mottled brown.

Common goldeneye,
American goldeneye

Bucephala clangula

SIZE: 18 inches

The male goldeneye identifies
himself by a distinctive white
spot on the side of the head
and large white wing patches.
The female lacks the white
spot; rather she has a brown
head and a yellow mark on the
end of her bill.

Canvasback

Aythya valisneria

SIZE: 19 to 24 inches

This is a light-colored duck.
The male has a red-brown
head and neck with a black
breast. The female is more
brownish in appearance. The
bill is black. The canvasback is
present in our area only in the
winter, as it nests in the
Northwest.

Redhead

Aythya americana

SIZE: 20 inches

This duck gets its common name from its most striking feature, the red head of the male. The female's head is more brownish. The male has a black chest and tail, gray back, and light underparts. The female is various shades of brown with a light area at the base of her bill. The bill is blue with a black tip.

Black duck

Anas rubripes

SIZE: 20 to 25 inches

The black duck has a mottled dark brown body and violet wing patches. The bill is olive-yellow, the feet red-brown.

Pintail

Anas acuta

SIZE: 25 to 30 inches

The male pintail's brown head with a white throat stripe and long, thin, pointed tail distinguish it from other ducks. The female, somewhat smaller, is an overall mottled brown. They both have green-bronze wing patches.

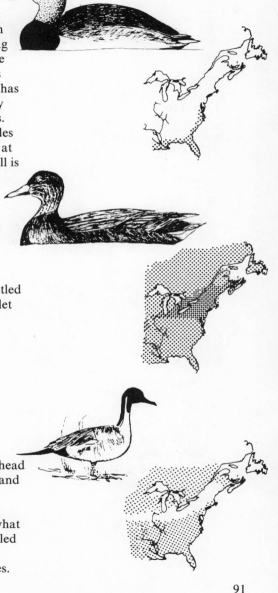

Wood duck

Aix sponsa

SIZE: 18 inches

The wood duck is unquestionably one of our most beautiful waterfowl. The male has a distinctive crest, iridescent green-and-purple head of metallic sheen with white stripes, spotted red-brown breast, and a conspicuous white bar in front of the wing. The female is gray-brown on the back with a white ring around the eye.

Ruddy duck

Oxyura jamaicensis

SIZE: 14 to 17 inches

O. jamaicensis is an interesting small bird. A white patch on the side of the head and a bristling, extended tail make the male quite distinctive. The female is rather plain in comparison.

American merganser,
common merganser

Mergus merganser

SIZE: 22 to 27 inches

This is the largest of the
mergansers. The male has a
green-black head, orange-pink
breast, white sides, and black
back. The female has a crested
red-brown head.

Swallow-tailed kite

Elanoides forficatus

SIZE: 21 inches

This is a chiefly southern bird
of prey. A fast, precise flyer, it
feeds primarily on insects and
small reptiles and amphibians.
It has a white head, black
body and wings, and long,
forked tail.

Marsh hawk

Circus cyaneus

SIZE: Up to 22 inches

The marsh hawk is various shades of brown, red-brown, and gray, with a tawny-white rump patch visible in flight. The male is generally grayer above. A close view shows a distinctive whitish ring about the face. *C. cyaneus* feeds primarily on small mammals and birds.

Osprey, fish hawk, fish eagle

Pandion haliaetus

SIZE: 21 to 24 inches

The osprey builds a large nest high up in trees or atop poles. A fish-eating hawk, it is common around lakes and ponds. It literally dives for its fish. The head and neck are white with a distinct black band through the cheeks. It has a dark brown back and barred wings and tail. The breast is white, with occasional brown streaks on the male, black spots on the female. The eyes are yellow, cere blue-gray, beak black, and feet bluish yellow. The sharp "elbows" of its wings in flight are an identifying characteristic.

American kestrel,
common kestrel, sparrow hawk

Falco sparverius

SIZE: 9 to 11 inches

This small, quick bird, with
the characteristic black mask
of a falcon, frequently hovers
in flight by means of rapid
wing beats. The male is
slate-blue above, with vertical
dark bars on the head, spotted
breast, white throat, and
amber-brown patches. The
female is less colorful, being
basically black-and-white with
brown bars on the tail and
without the bluish hue on her
back. The beak is blue-black;
cere, feet, and legs are
yellow-orange.

King rail

Rallus elegans

SIZE: 16 to 18 inches

To identify the king rail, look
for a long, slender bill, a
red-brown back streaked with
black, and cinnamon cheeks
and breast.

Florida gallinule,
common gallinule

   Gallinula chloropus

SIZE: 13 to 14 inches

The shield on the forehead is
unusual enough to identify
this bird. It has a red bill, a
gray head, neck, and breast,
an olive back, and white
feathers on the flanks.

American coot, common coot

   Fulica americana

SIZE: Up to 16 inches

The coot is generally gray with
a darker head and neck. Its
bill is nearly white. It has
green legs and shows white
under the tail feathers.

Killdeer

   Charadrius vociferus

SIZE: 10 inches

Two prominent black breast
bands distinguish the killdeer
from other ground nesters. It
has a white forehead, a black
bill, a brown crown, back, and
tail, and white underparts.

96

Wilson's snipe, common snipe

Capella gallinago

SIZE: 11 inches

At a distance its chunky appearance, long, slender bill, and cry of "whee-whee" distinguish the snipe from other marsh birds. It is striped brown-and-white above, with a spotted breast, a small orange tail, and brown-green feet.

Solitary sandpiper

Tringa solitaria

SIZE: Up to 9 inches

A dark bird, the solitary sandpiper is best distinguished by the white feathers on the sides of its tail, which are clearly visible in flight. It is generally seen only during migration because of its great variation in range.

Spotted sandpiper

Actitis macularia

SIZE: Up to 8 inches

This is our most common
sandpiper. It is identified in
the spring by the spots on its
breast. It has an olive-brown
back, a gray-white breast, and
a white line over the eye.

spring plumage

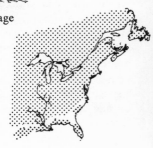

Greater yellowlegs

Totanus melanoleucus

SIZE: 14 inches

As the name implies, its long,
stiltlike yellow legs are this
bird's best field mark. It is
dark gray-brown with streaks
on the neck and chest, and has
a slightly upturned dark bill.

Common kingfisher,
belted kingfisher

Megaceryle alcyon

SIZE: 11 to 15 inches

A crested gray-blue bird which
dives into the water is
unmistakably the kingfisher. It
has a wide white collar and
blue breast band. Red-brown
marks are seen on the belly
and flanks.

Tufted titmouse

Parus bicolor

SIZE: 6 inches

The tufted titmouse, a close
cousin of the chickadee,
distinguishes itself as our only
small gray songbird with a
distinctive crest. Its song is
similar to that of the
chickadee, except perhaps a
bit deeper. It has a white
breast and a red-brown patch
on its flanks.

99

Long-billed marsh wren

Telmatodytes palustris

SIZE: 5 inches

A small brown bird with a
white line over the eye, darting
about in cattails, is bound to
be the long-billed marsh wren.
A close look would reveal
black-and-white stripes on the
back. It is difficult to find as it
streaks away at the slightest
disturbance.

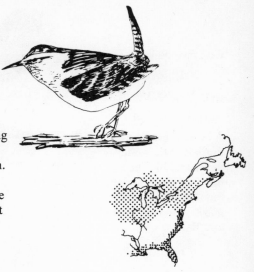

Yellow warbler

Dendroica petechia

SIZE: 5 inches

The yellow warbler is easy to
identify as it is the only small
bird in our range that appears
yellow from any direction. A
closer look shows red-brown
streaks on the male's breast;
these are very faint or lacking
entirely on the female.

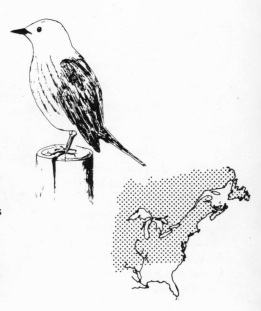

Redwing blackbird

Agelaius phoeniceus

SIZE: 9 inches

The bright red-and-yellow
flash of the male blackbird's
shoulder bars is as well known
as the cattails and rushes it
nests among. The female is
less distinctive, being a rather
dull brown for camouflage in
the marsh vegetation.

Swamp sparrow

Melospiza georgiana

SIZE: 6 inches

Most sparrows are look-alikes,
and this one is no exception. It
is a bit chunkier than others
and has plain gray underparts.
There are no wing bars. Look
for a reddish crown, a white
throat, red-brown wings, and a
rounded tail.

Deer mouse

Peromyscus maniculatus

SIZE: Body 3 to 4 inches;
tail 3 to 5 inches

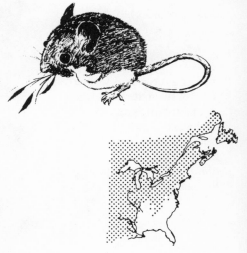

This attractive rodent is
blue-gray as a youngster, and
varies from red-brown to
gray-brown as an adult.
Underparts are nearly white.
A marsh provides excellent
forage for the deer mouse,
whose diet consists of fruits,
nuts, seeds, insects, and other
small invertebrates. Deer mice
start breeding at seven weeks
of age; a litter averages four.

Bog lemming,
southern bog lemming,
lemming mouse

Synaptomys cooperi

SIZE: 3½ to 4½ inches

The bog lemming is a strange
creature indeed. One year they
will be seen scurrying about;
another year there won't be a
sign of them. This animal
prefers cold-water wetlands,
where it travels quickly
through the vegetation or
moves through tunnels it has
dug. Capable of reproducing
before fully grown, it breeds
from March to October, with
an average of three per litter.
The bog lemming is a grizzly
brown, lighter underneath.

Norway rat

Rattus norvegicus

SIZE: Body 7 to 10 inches;
tail to 8 inches

The gray-brown Norway rat
has earned the dislike of
people everywhere through its
destructive behavior and habit
of spreading disease. It is a
vicious and dangerous biter if
handled. Its life span is about
four years.

Black rat

Rattus rattus

SIZE: Body to 8 inches;
tail over 8 inches

The black rat is commonest in
the southern part of our range.
It is distinguished from the
Norway rat by its slender
body and longer tail. It
disdains little in the way of
food, subsisting on nearly
anything available. There are
five to ten young in a litter;
several litters may be born in a
year.

Muskrat

Ondatra zibethicus

SIZE: Body 10 to 15 inches;
tail 9 to 12 inches

The muskrat is a familiar
marsh inhabitant except in the
southeastern part of our range.
Its dark brown fur has been
sought by trappers since early
history. The tail is vertically
flattened, the hind feet partly
webbed. The muskrat builds a
stick house similar to a
beaver's, with an underwater
entrance. It is active both day
and night. As many as ten
young (usually five to seven)
are born several times
annually; they are weaned in
about a month.

Beaver

Castor canadensis

SIZE: Body 20 to 30 inches;
tail 7 to 10 inches

The beaver, like the muskrat,
is widely sought for its fur.
Legislation controlling its
catch saved the species from
annihilation. The dense dark
brown fur protects it from
ice-cold water, and the flat,

104

scaled tail helps propel it in swimming. Beavers build "lodges" in the marsh out of sticks and branches; the lodge is provided with an underwater entrance. They will construct a dam on a stream to flood a lowland and provide themselves a suitable environment. The beaver feeds on the bark, twigs, leaves, shrubs, and roots of deciduous trees. A litter comprising two to four young is born in April, May, or June. The young stay with their family for two to three years.

Eastern gray squirrel

Sciurus carolinensis

SIZE: Body 7 to 10 inches; tail 7 to 10 inches

The gray squirrel is not truly a marsh animal, but it is not a rare instance to see it scurrying about in the wetlands. The great variety and quantity of vegetation there is quite an attraction to such animals. *S. carolinensis* has a gray body and tail; the tail fur is white-tipped. A black phase is occasionally encountered in the northern part of its range.

Water shrew

Sorex palustris

SIZE: Body 3 inches;
    tail 2 inches

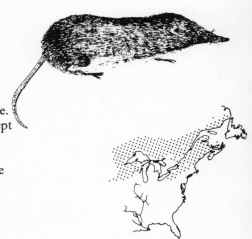

The shrew is black or nearly
black with a lighter underside.
An insect eater, it is very adept
in the water. Between April
and October one or more
litters of two to ten young are
born. The young are capable
of fending for themselves in
about three weeks.

Raccoon

Procyon lotor

SIZE: Up to 35 inches

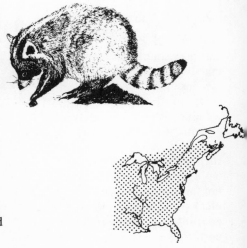

The raccoon is not easily
confused with any other
animal because of its
distinctive black face mask
and ringed tail. It is most
active at night, when it
searches out its food. The
marsh is an abundant
environment for the raccoon.
It disdains little, eating most
vegetables and fruit, birds and
bird eggs, fish and crawfish.

Virginia deer,
white-tailed deer

Odocoileus virginianus

SIZE: Up to 6 feet long;
3½ feet high at shoulder

Deer are attracted to the
wetlands by their rich supply
of grass, leaves, and shrubs. In
the summer the "white-tail" 's
coat is red-brown, becoming
grayer in the winter. The
underside of the tail is brilliant
white. The fawn has white
spots on red-brown fur. There
are two fawns born in the
summer.

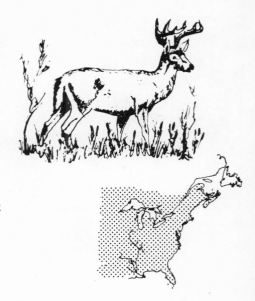

Moose

Alces alces

SIZE: Up to 10 feet long;
up to 8 feet high at shoulder

It is impossible to confuse this
animal with any other. Its
large, palmated antlers, huge
snout, and massive body
distinguish the moose. A good
swimmer, it feeds on aquatic
plants as well as terrestrial
vegetation. In May or June
one or two calves are born.

107

BIBLIOGRAPHY

Fernald, M. L., *Gray's Manual of Botany* (eighth ed.). Van Nostrand Reinhold, New York, 1970.

Niering, W. A., *The Life of the Marsh*. McGraw-Hill, New York, 1966.

Petrides, G. A., *A Field Guide to Trees and Shrubs*. Houghton Mifflin, Boston, 1958.

Peterson, R. T., and M. McKenny, *A Field Guide to Wildflowers*. Houghton Mifflin, Boston, 1968.

Snedigar, R., *Our Small Native Animals: Their Habitats and Care*. Dover, New York, 1963.

Zim, H., and A. Martin, *Flowers: A Guide to Familiar American Wildflowers*. Simon & Schuster, New York, 1950.

Raven, P. H., and H. Curtis, *Biology of Plants*. World, New York, 1970.

Stanek, V. J., *The Pictorial Encyclopedia of Insects*. Hamlyn, London, 1969.

Swan, L. A., and C. S. Papp, *The Common Insects of North America*. Harper & Row, New York, 1972.

Collins, H. H., *Complete Field Guide to American Wildlife*. Harper, New York, 1959.

Nichols, D., J. Cooke, and D. Whiteley, *The Oxford Book of Invertebrates*. Oxford, 1971.

Ursin, M. J., *Life In and Around the Salt Marshes*. Crowell, New York, 1972.

Newman, J. R., ed., *The Harper Encyclopedia of Science*, 2 vols. (rev. ed.). Harper & Row, New York, 1967.

Peterson, R. T., and J. Fisher, *The World of Birds*. Doubleday, Garden City, N.Y., 1964.

Grossman, M. L., and J. Hamlet, *Birds of Prey of the World*. Potter, New York, 1964.

Borror, D. J., and R. E. White, *A Field Guide to the Insects of America North of Mexico.* Houghton Mifflin, Boston, 1970.

Rothschild, Lord, *A Classification of Living Animals.* University Press, Glasgow, 1962.

Hvass, H., *Birds of the World.* Dutton, New York, 1963.

Klots, E. B., *The New Field Book of Freshwater Life.* Putnam, New York, 1966.

Hotchkiss, N., *Common Marsh, Underwater and Floating-Leaved Plants of the United States and Canada.* Dover, New York, 1972.

Leviton, A. E., *Reptiles and Amphibians of North America.* Doubleday, Garden City, N.Y., 1972.

Wetmore, A., *Water, Prey and Game Birds of North America.* National Geographic Society, Washington, D.C., 1965.

Wild Animals of North America. National Geographic Society, Washington, D.C., 1960.

Wetmore, A., *Song and Garden Birds of North America.* National Geographic Society, Washington, D.C., 1960.

Went, F. W., et al., *The Plants.* Time-Life, New York, 1963.

Wright, A. H., and A. A. Wright, *Handbook of Snakes.* Cornell, Ithaca, N.Y., 1957.

Walker, E. P., *Mammals of the World.* Johns Hopkins, Baltimore, 1964.

The Larousse Encyclopedia of Animal Life. McGraw-Hill, New York, 1967.

Buchsbaum, R., and L. J. Milne, *The Lower Animals: Living Invertebrates of the World.* Doubleday, Garden City, N.Y., 1967.

Gillette, J. D., *The Mosquito.* Doubleday, Garden City, N.Y., 1972.

Matthiesson, P., *Wildlife in America.* Viking, New York, 1964.

INDEX

dragonfly, 63
Dryopteris intermedia, 21
Dryopteris sp., 21
Dugesia dorotocephala, 55
Dytiscus fasciventris, 66

eastern chain fern, 22
eastern crane fly, 70–71
eastern gray squirrel, 105
eastern newt, 76–77
eastern red cedar, 52
eastern spadefoot, 80
Echinochloa crusgalli, 31
Echinodorus cordifolius, 25
Eichhornia crassipes, 38
Elanoides forficatus, 93
Eleocharis acicularis, 32
Eleocharis elongata, 31
Eleocharis obtusa, 32
Eleocharis rostellata, 31
Elliptio crassidens, 58
Emys blandingi, 81
Equisetum fluviatile, 20
Euglena sp., 15
euglenoids, 15
EUGLENOPHYTA, 15
Eurycea bislineata, 76
everglade beakrush, 35
Eylais sp., 73

Falco sparverius, 95
false longhorn leaf beetle, 66
fancy fern, 21
fasciated diving beetle, 68
ferns, 20–22
Filicineae, 20–22
fish eagle, 94
fisher spider, 74
fish hawk, 94
flagellates, 14
flagroot, 37
flatworms, 55
floating water primrose, 46
flood control, 4
Florida gallinule, 96
flowering plants, 22–51
flowering rush, 26
folktales, wetlands in, 3–4
Fowler's toad, 80
Fragillaria sp., 13
freshwater cordgrass, 28
freshwater jellyfish, 54
Fulica americana, 96

Gallinula chloropus, 96
Gammarus sp., 71

Gelastocoris oculatus, 59
Gerris sp., 61
giant cutgrass, 29
giant pond snail, 57
giant water bug, 59
glaciers, 1
Glyceria acutiflora, 27
golden club, 37
Graptemys geographica, 81
gray treefrog, 77
great blue heron, 86–87
greater yellowlegs, 98
green algae, 17–18
green frog, 79
green heron, 87
Gyraulus sp., 57

hairy smartweed, 41
hairy wheel snail, 57
hardstem bulrush, 33
hellbender, 74–75
hellgrammite, 65
Heteranthera reniformis, 39
Hexagenia sp., 62
Hibiscus moscheutos, 45
Hippuris vulgaris, 47
horsetails, 20
huntsman's cup, 44
hydras, 54
Hydra sp., 54
Hydrochus scabratus, 68
Hydrometra martini, 62
Hyla crucifer, 78
Hyla versicolor, 77

Indian cup, 44
Insecta, 59–71
insects, 59–71

jellyfishes, 54
jointgrass, 30
Juncus balticus var. littoralis, 40
Juncus effusus var. solutus, 39
Juncus militaris, 40
Juncus pelocarpus, 41
Juncus repens, 40
Juniperus virginiana, 52
Jussiaea repens, 46
Justicia americana, 49

killdeer, 96
king rail, 95

Laccophilus fasciatus, 68
lake sedge, 36
larch, 51

swamp, definition of, 2
swamp bulrush, 34
swamp chorus frog, 77
swamp cricket frog, 77
swamp lily, 50
swamp loosestrife, 45
swamp smartweed, 43
swamp sparrow, 101
sweetflag, 37
Symplocarpus foetidus, 26
Synaptomys cooperi, 102–103

tamarack, 51
Telmatodytes palustris, 100
Thelypteris palustris, 21
Tipula abdominalis, 70–71
toad bug, 59
Totanus melanoleucus, 98
TRACHEOPHYTA, 20–52
Trichoptera, 65
Triglochin maritima, 24
Tringa solitaria, 97
Trionyx ferox, 83
Trionyx muticus, 82
tufted titmouse, 99
twig rush, 35
two-lined salamander, 76
Typha angustifolia, 23
Typha latifolia, 22

vascular plants, 20–52
VERTEBRATA, 74–107
vertebrates, 74–107
Virginia chain fern, 22
Virginia deer, 107
Viviparus intertextus, 57

walking spikerush, 31
water beech, 50
water boatman, 60

water feather, 46
water flea, 71
water horsetail, 20
water hyacinth, 38
water milfoil, 46
water mite, 73
water moccasin, 84
water moss, 18
water parsnip, 48
water pipes, 20
water purslane, 46
water scavenger beetle, 68
water scorpion, 61
water shrew, 106
water skater, 61
water smartweed, 42
water spikerush, 31
water strider, 61
water willow, 45, 49
wetlands:
    destruction of, 4
    distribution of, 9
    origins of, 1–2
    and public attitude, 3
    types of, 2
whirligig beetle, 67
whistling swan, 88
white-tailed deer, 107
wild millet, 31
wildrice, 29
Wilson's snipe, 97
winkle, 57
wood duck, 92
Woodwardia virginica, 22

yellow pond lily, 43
yellow warbler, 100

Zizania aquatica, 29
Zizaniopsis miliacea, 29
Zygoptera, 64

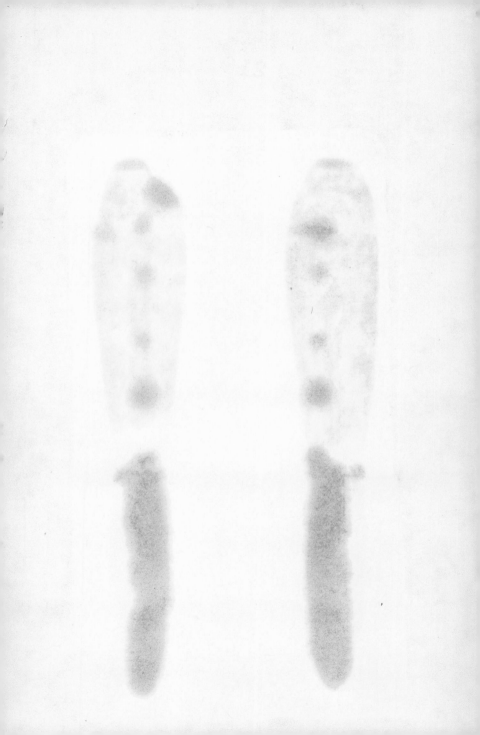